T0146964

RECONCILING
GENESIS & SCIENCE

UNLOCKING THE THEORIES OF CREATION

FRED SNOWDEN

authorHOUSE®

AuthorHouse™
1663 Liberty Drive
Bloomington, IN 47403
www.authorhouse.com
Phone: 1 (800) 839-8640

Published by AuthorHouse 10/09/2019

ISBN: 978-1-7283-2941-3 (sc)
ISBN: 978-1-7283-2939-0 (hc)
ISBN: 978-1-7283-2940-6 (e)

Library of Congress Control Number: 2019915250

CONTENTS

PREFACE

I have spent much of my life and career mulling over the wonders of creation. All of us innately wonder about the existence of the universe and of ourselves. Today, as with politics, Americans are divided over the issues of evolution and creationism, but few ever stop to thoroughly study these concepts. My purpose is to put the most convincing theories from all perspectives in one place and let the readers decide. Hopefully, what makes sense and what doesn't will become clearer. More important, my hopes are that proponents from both sides will relax and be less adversarial. It is understood that the questions surrounding the creation of the universe are significant, but not to the point of making enemies of those who are only seeking truth. Hopefully, an understanding of our opponents' motivations and attitudes may create some degree of empathy and compassion. It might even be possible for people with opposing views to become colleagues and friends.

Have you ever heard that modern humans may have existed for over a million years?

Are you aware that there might be a vast difference between the orbital and atomic calendars?

Did you know modern research suggests that we are less like chimpanzees than first suggested?

Can you believe that there is overwhelming evidence that the human brain is still evolving?

Have you considered that the mighty Zulus and the tiny pygmies in Africa prove Darwin's theory of evolution?

This and much more awaits you, but first let's see why we believe what we believe about evolution and creationism.

CHAPTER 1

SEEKING ANSWERS FROM PHILOSOPHY

The modern controversy regarding creation has become an out-of-control social phenomena. The conflict is driven by passion and fueled by prejudice. The battlefields include schools, bookstores, libraries, classrooms, museums, churches, mosques, temples, and synagogues. The winners are those who profit from the war. The losers are school-age students, sincere college students, and the rest of us who want factual information rather than defensive rhetoric, objectivity as opposed to blind passion, and transparency rather than a strategic cover-up. The shame is that there are sincere, intelligent people on both sides of science and faith, but most are so caught up in defending their sacred ground that they will not even listen to, much less consider, each other's viewpoints. When the battle lines are finally drawn, the caissons line up, each with a different worldview to protect and defend using the ammunition they hold.

On the extreme left are the secularists or objectivists who believe that the universe and all life is self-perpetuating through natural law; they reject the idea of any intelligence or design behind creation. To them, the Genesis narrative is primitive nonsense, comparable to *The Lord of the Rings* or "Jack and the Bean Stalk." Objectivism holds that reality is an absolute regardless of one's fears, dreams, hopes, or desires. It holds that ideas must be true and thus practical. Objectivism rejects that reality is determined by personal opinion or social convention, so an individual's ideas or beliefs do not make reality what it is. Therefore, there is no evidence or rational argument to support the existence of a supernatural being who creates and controls reality.

On the extreme right are the young Earth creationists who believe that God created a fully matured universe about six thousand years ago in six regular, twenty-four-hour days, with the light from the farthest stars already visible, exactly as described in the Genesis narrative. This position may also be called literalism, which is the adherence to the explicit substance of an idea or expression and is most associated with biblical interpretation. It equates to the dictionary definition of literalism, "adherence to the letter, word and literal sense, where literal means the strict sense of the word or words, not figurative or metaphorical."

To the right of the objectivists we find the theistic evolutionists, or compatibilists, who believe that the hand of the Creator initiated and guided the processes of creation and evolution over eons of time. In their minds, the Big Bang and Darwinian evolution are simply tools on the tool belt of the Creator that have produced what we observe today. For them, the Genesis narrative is but an understandable explanation of a highly complex operation from the people in the day that the Bible was written, given their limited understanding of science. Compatibilism offers a solution to the free will problem because it holds that free will is compatible with determinism, in that free will is normally taken to be a necessary condition of moral responsibility.

Another step to the right takes us to intelligent Design (ID) creationists, who believe that an intelligence created the universe, the Earth, and each individually specific life form. These "scientists" are careful to stay away from words like *God* or *Bible* so that they appear to be guided by information rather than religion. As is often pointed out by objectivists, there is a close relationship between intelligent design theory (IDT) and traditional creationism. Most intelligent design theorists believe in the old Earth theory and accept some degree of overall common descent. The Discovery Institute, a privately-supported think tank in Seattle, funds much of the work of the IDT movement. The driving force at the institute is University of Chicago–educated young Earth creationist and philosopher Paul Nelson.

Between ID creationists and the literalists, we find the first cousins of the young Earth creationists (YEC): traditional creationists or theists. They believe that the ID's intelligence behind creation is the God of the Bible but that the six "days" of the Genesis narrative are six eons of immeasurable time, not twenty-four-hour days as the literalists insist. They would say that they are the literalists and that their right-wing, young Earth cousins are actually hyperliteralists. Their belief in an old Earth aligns them with objectivists and theistic evolutionists when it comes to geology, carbon dating, and more. Holding that reality has an inherently logical structure, the theists asserts that truths exist, which the intellect can grasp directly without going to extremes. These rational principles in logic, ethics, and even metaphysics are so fundamental that to deny them is to fall into contradiction. For traditional creationists, Darwinian evolution and YEC are two of those contradictions.

One might wonder how could there be so many opinions about one issue held by so many distinguished scholars of science, theology, and philosophy. The simple answer is that in the case of the creation, the first and most important step in the scientific method, observation, is missing. No one observed the actual creation. Without that foundational step, there is and will continue to be controversy regarding the process. You see, abstract ideas, opinions, and personal perspective are a product of many mitigating factors, which begin in the individual mind at the molecular level. Thus, all human ideas are fueled by philosophy and worldview— what one believes about life. Philosophy answers the big questions.

Am I a cosmic accident or a purposely designed individual?

Where did the human soul come from?

Why am I here?

What is my purpose?

Everyone has a philosophy or worldview, even if one doesn't recognize it or cannot express it in specific terms. Our philosophy or worldview reveals itself as we live our lives, develop our values and mores, and raise our children. The ancient Greeks understood philosophy better than any other culture and left to us untold volumes recognizing tendencies and ascribing descriptive names and characteristics. Very few people today can say whether they are a Stoic or an Epicurean, but the trained eye can easily discern this from one's lifestyle and values.

Worldview plays a major role in choosing a creation belief. Objectivists, who are atheists or agnostics, have no choice but to reject the Genesis narrative, whereas most Christians, Orthodox Jews, and Muslims have no choice but to accept it. A prime example is Dr. Francis Collins, the primary investigator and brilliant mind who unlocked of the human genome. For much of his life, Dr. Collins was an avowed atheist and Darwinian evolutionist, yet later in life he became a Christian. Over time, as he studied the Bible and grew in faith, his worldview drastically changed. Faith in God and an understanding of the scriptures realigned what he had previously believed about creation. In order to accommodate both faith and evolution, he compatabilized that theistic evolution—evolution guided by the hand of the Creator, not random chance—was the best explanation for what he had observed in nature and the laboratory. In this particular creation belief, he was able to retain what he understood about evolution and was not forced to surrender his scientific objectivity or intellectual integrity.

Rational philosophy can only lead to four propositions regarding the Genesis narrative, and these have been held through the ages in various forms.

- The Genesis narrative is a fable, there is no supernatural force in creation, and everything that exists is a product of the force of nature at work (evolutionist, objectivist).
- The Genesis narrative is an allegorical, creation myth, which presents an understandable, unscientific explanation for the unlearned (theistic evolutionist, compatibilist).
- The Genesis narrative is superfluous because science demonstrates that there is intelligence behind creation and that evolution is contradictory and irrational (ID creationist).
- The biblical narrative is a literal, reasonable, scientific explanation for what we observe in nature (creationist, literalist).

Before moving on, it is important to distinguish between the terms *belief* and *knowledge*. Belief may be abstract and unsupported by fact or reason; it is mostly associate with religion, is respected and even protected, but is not held as truth. *Merriam-Webster's Dictionary* defines belief as "a feeling of being sure that something exists or is true, right or valuable." Knowledge is on a different level than belief. It is held as science supported by a precise set of facts and observations. All of the positions previously listed would say that their understanding of creation is based in supportable fact and is thus is knowledge, not simply belief. For this reason, each must be held accountable to a higher standard. The intent of this book is to scrutinize the facts to which each group clings in an effort to sort out fact from fiction, knowledge from belief, and truth from fable.

Let us examine these positions, beginning with the domain of objectivism. It includes all those who never consider the possibility of a creator. For them, there is a naturalistic explanation for creation, so they must develop a scientific model that is completely independent of the supernatural. Objectivists are typified as unbridled liberals, but Ayn Rand, a social and economic conservative, defined her philosophy as objectivism and called it "a philosophy for living on earth." Objectivism is an integrated system of thought that defines the abstract principles by which a human must think and act if one is to live the life proper to humankind. Ayn Rand first portrayed her philosophy in the form of the heroes of her best-selling novels, *The Fountainhead* (1943) and *Atlas Shrugged* (1957). She later expressed her philosophy in nonfiction form. She was once asked if she could present the essence of Objectivism in a nutshell. Her answer was as follows.

Metaphysics:	Reality
Epistemology:	Reason
Ethics:	Self-interest
Politics:	Capitalism

She then translated those terms into familiar language.

"Nature, to be commanded, must be obeyed."
"You can't eat your cake and have it, too."
"Man is an end in himself."
"Give me liberty or give me death."

For the objectivist, reality, evidenced by the external world, exists independent of humankind's consciousness, independent of any observer's knowledge, beliefs, feelings, desires, or fears. This means that two plus two always equals four, facts are facts, beliefs are invaluable, things are what they are, and the task of a human's consciousness is to perceive reality and not to create or invent it. Thus, objectivism rejects any belief in the supernatural and any claim that individuals or groups can create their own reality. Objectivism also rejects mysticism, the idea that any acceptance of faith or feeling is a means of knowledge, and skepticism, the claim that certainty or knowledge is impossible. Humans are rational beings, and reason, as humankind's only means of knowledge, is the basic means of survival. However, the exercise of reason depends on each individual's choice. This one choice controls all the other choices you make and determines your life and character. Objectivism also rejects any form of determinism—the belief that man is a victim of forces beyond his control, such as God, fate, upbringing, genes, or economic conditions. Reason is humankind's only proper judge of values and the only proper guide to action. The proper standard of ethics is that which is required by humankind's nature for survival as rational beings rather than as mindless brutes.

Another form of objectivism is skepticism, which refuses to make any claims of truth, although a philosophical skeptic does not entertain that knowing truth is impossible. The label is commonly used to describe other similar philosophies such as academic skepticism, an ancient variant of Platonism that claimed knowledge of truth was impossible. Philosophical skepticism originated in ancient Greek philosophy.

Compatibilists, exemplified in the previously mentioned Francis Collins, rationally examine the facts and see an intelligent, guiding hand or supernatural force in creation. Thus, they are convinced that true science supports such a position even if the specifics of that force cannot be defined. Hume explained that the compatibilists' free will should not be understood as some kind of ability to have actually chosen differently in an identical situation. The compatibilist believes that a person always makes the only decision possible; any talk of alternatives is strictly hypothetical. Faced with the standard argument against free will, many compatibilists choose determinism so that their actions are adequately determined by their philosophy, reasons, motives, culture, and more. Objectivists (also described as hard determinists) accuse compatibilists (also described as soft determinists) of being motivated by a lack of a coherent, consonant moral belief system.

William James accused them of creating a "quagmire of evasion" by stealing the name of freedom to mask their underlying determinism. Immanuel Kant called it a "wretched subterfuge" and "word jugglery." Kant's argument turns on the view that although all empirical phenomena must result from determining causes, human thought introduces something seemingly not found elsewhere in nature: the ability to conceive of the world in terms of how it ought to be, or how it might otherwise be. Because of its capacity to distinguish *is* from *ought*, reasoning can spontaneously originate new events without being itself determined by what already exists. Kant proposes that to take the compatibilist view is to deny the distinctly subjective capacity to rethink an intended course of action in terms of what ought to happen.

Literalists, or theists, are certain that specific supernatural revelation (such as the Tora, Quran, or Bible) provides the only trustworthy explanation for creation. Most evangelical Christians, Orthodox Jews, and Muslims, and some Hindu sects, are creationists in this sense. The creationist believes in order, design, and therefore an ultimate designer. For them, this view has been constantly borne out by the new discoveries of the last century, of which the genome is just one. This intricate blueprint that is unique to each human individual and indeed is the makeup of all living organisms could not have developed by chance or accident. Therefore, a creationist is one who trusts in God rather than the proverbial primeval soup as the creator of all life on earth.

Creationists take comfort in the accomplishment of miracles by a higher power. They believe that God is omnipotent, so evolution in any form cannot be valid. Evolution is but an unproven theory that is subject to constant revision, rendering it almost laughable. They are completely comfortable with facing the obvious conflicts held to be "modern reason" and are willing to be described as flat-earthers by those who reject the idea of a supernatural creation because they understand how it is motivated. Most creationists are also castrophists and depend on other supernatural occurrences, like those mentioned in the Bible, to support their claims, especially the global deluge. They take confidence that such an event is also mentioned in extra-biblical literature of the ancient world. Finally, many creationists see a strong link between evolution and hotly contested current moral issues: abortion on demand, animal rights, and homosexuality as a genetic predisposition. <any literal, six-day creationists are also dispensationalists, so this creation model best fits their view of eschatology that God is working through several distinct periods of time to complete the history of mankind. C. S. Lewis summed up the philosophy for theism in his essay "Is Theology Poetry?"

> One absolutely central inconsistency ruins the popular scientific philosophy. The whole picture professes to depend on inferences from observed facts. Unless inference is valid, the whole picture disappears … unless Reason is an absolute, all is in ruins. Yet those who ask me to believe this world picture also ask me to believe that Reason is simply

the unforeseen and unintended by-product of mindless matter at one stage of its endless and aimless becoming. Here is flat contradiction. They ask me at the same moment to accept a conclusion and to discredit the only testimony on which that conclusion can be based.

This argument against naturalism and materialism holds the following.

1. No belief is rationally inferred if it can be fully explained in terms of nonrational causes.
2. If naturalism is true, then all beliefs can be fully explained in terms of nonrational causes. In other words, they can be explained by factors in nature, such as the four governing forces.
3. Therefore if naturalism is true, then no belief is rationally inferred.
4. If any thesis entails the conclusion that no belief is rationally inferred, then it should be rejected and its denial accepted.

Conclusion: Therefore, naturalism should be rejected and its denial accepted.

The argument for the existence of a creator is as follows.

- A being requires a rational process to assess the truth or falsehood of a claim (hereinafter, to be convinced by argument).
- Therefore, if humans are able to be convinced by argument, their reasoning processes must have a rational source.
- If humans can be convinced by argument, their reasoning processes must have a nonphysical as well as a rational source.
- Rationality cannot arise out of nonrationality. That is, no arrangement of nonrational materials creates a rational thing.
- No being that begins to exist can be rational except through reliance, ultimately, on a rational being that did not begin to exist. That is, rationality does not arise spontaneously from out of nothing but only from another rationality.
- All humans began to exist at some point in time.

- Therefore, if humans are able to be convinced by argument, there must be a necessary and rational being on which their rationality ultimately relies.
- Conclusion: This being we call God.

Islamic scholar Hamza Andreas Tzortzis put it this way.

> No question is more sublime than why there is a universe: why there is anything rather than nothing. When we reflect upon our own existence we will come to the realization, that at some point in time, we began to exist. Since we were once non-existent and are now in existence, it follows that we must have had a beginning. In light of this, the Qur'an raises some profound questions.

Quran 52:35–36 states, "Were we created by nothing? Did we create ourselves? Or did we create the universe? Or were they created by nothing? Or were they the creators (of themselves)? Or did they create heavens and earth? Rather, they are not certain."

Objectivists would say that creationists, through a process known as cognitive dissonance (the state of holding two or more conflicting ideas, beliefs, values and/or emotional reactions), accommodate reality and achieve a philosophical balance or psychological awareness of uncomfortable tension caused by holding two conflicting thoughts or beliefs in one mind at the same time. Because creationists tend to seek consistency in their beliefs and perceptions, something must change in order to eliminate or reduce the dissonance. The theory of cognitive dissonance in social psychology proposes that people have a motivational drive to reduce dissonance by altering existing cognitions, adding new ones to create a consistent belief system, or reducing the importance of any one of the dissonant elements.

On the other hand, creationists might say that it is the objectivists who allow cognitive dissonance to comfort them in their delusion. It's a delusion because they are anthropologically aware of the existence of a creator but reject him in order to escape a life lived in submission to such an invisible higher power. In the absence of genuine proof of life arising from nonliving matter, of undisputable transitional forms in the fossil record, and in the face of the overwhelming mathematical improbability that humans could evolve from lower forms, objectivists cling to their "faith" in nature and blindly refuse to consider that there could be intelligence behind creation.

If we agree that one's worldview or philosophy is the driving factor in what one believes about creation, we can also agree that worldview creates more questions than answers. Philosophy is totally subjective, so it seems as if our creation beliefs are simply a reinforcement of our philosophical bias. Thus the contempt, vehemence, and downright subterfuge that take place in the schools, media, and laboratories are fueled more by philosophy than pure science, and the combatants can never surrender. The validated facts that are held in common by all sides are interpreted with particular philosophical bias. If they do not have good evidence, they fake it until they do. It is always too soon or too late to tell. If they fail, they cover it up and dig it up again in at the same spot, hoping for a new twist. Intelligent scientists, philosophers, and theologians are thus reduced to dogs chasing their own tails.

So what are the seekers of truth to do? We must make a sincere attempt to sort out fact from fiction, science from theory, and sound theology from blind faith. There is room for debate, and there is room for compromise. However, there is no longer room for the soapbox. It is time for a truce in which all combatants stop bickering and start listening to one another in an effort to generate a genuine, empathetic understanding of each other's points of view. Agreement is not a realistic goal, but a willingness to admit that a person's reality is not flawed simply because he or she does not agree about an event that was not observed is.

A rational, unvarnished understanding of the scientific method is critical to exploring the subjects and issues that will be discussed hereafter. The elements of this essential system are as follows.

- ➢ Observation—gathering data
  - ➢ Looking at nature
  - ➢ Quantifying and organizing data
  - ➢ Researching what others have observed

➢ Hypothesis/theory—formulating a reasonable explanation
  ➢ Must be testable (practical)
  ➢ Must be reproducible over a period of time
➢ Experimentation—an artificial, controlled investigation
  ➢ Must have a constant (that which does not change)
  ➢ Must have a variable (that which changes)
➢ Only after a hypothesis is proven through many experiments and has the endorsement of other scientists does it become a theory
➢ When a theory endures the test of time, it becomes scientific law
➢ There have been occasions when accepted scientific laws have been redefined, as with Bode's Law

Bode was a great mathematician, astronomer, and physicist who was hindered by the limitations of the telescopes available in his day. He observed certain patterns in the orbits of the planets and mathematically quantified these observations. In applying his law, he predicted that there must be a hidden planet between Mars and Jupiter. His hypothesis made reasonable sense and was investigated by other scientists, which elevated his prediction to a theory. When the asteroid belt was discovered, the scientists of that day thought that they had proven Bode's theory, and it became scientific law. With improvement to the telescope, however, Saturn was discovered, as was Neptune, which better explained Bode's predictions, and his prediction of the hidden planet was dismissed. This does not negate his genius, but it does make a good point: even established scientific law may be open to modification as better instruments and new discoveries give rise to new evidence. This brings us to the limitations of science.

▸ Science is limited to proving predictions and theories related to matter and energy (the physical universe) that exist in the present
▸ Things that are spiritual, supernatural, moral, or mystical cannot be quantified; these belong to the philosophy of metaphysics
▸ Things that occurred in the past and that might occur in the future cannot be observed or tested, so for now, theories or predictions dependent upon them must remain unproven
▸ Artifacts, archeological digs, and peering into deep space give us clues to the past, but certain assumptions that are not science

may still be required, even if those assumptions are held for long periods of time

▸ Although philosophical or worldview bias should not affect or scientific objectivity, it often does

>We often accept theory as fact
>We often accept long held assumptions as fact
>We often accept what cannot possibly be proven as fact

CHAPTER 2

SEEKING ANSWERS FROM THE BEGINNING

The first words in the Bible are "In the beginning." Both creationists and evolutionists have long asked, "When was that beginning? At what point did the envelope which we call time and space come into being? When did matter, energy, and the four essential forces make their appearance?" Bishop Ussher fixed the creation date as 4004 BC by using generational records from scripture, but most physicists and astronomers today argue that the universe is between 13.4 and 13.8 billion years old.

Astronomers and physicists have become very dogmatic about the accuracy of their timing of the beginning of the universe since the discovery of the red shift in space, which, as far as most are concerned, proved the theory of an expanding universe. Starting with Stephen Hawking and based on recent discoveries by the Hubble Telescope, scientists claim to have been able to reverse the expansion of the universe from the present to the moment at which the universe came into existence from nothing. They say that they now have the tools and ability to, in a manner of speaking, rewind the creation event from the present to the beginning. It is somewhat like rewinding a movie on a DVD. When we push the rewind button, we are able go from the end to the beginning and observe the images in the order things actually happened. Scientists contend that what has been discovered through this process is equivalent to observation itself.

This is not the only evidence that modern objectivists rely upon to estimate the age of the universe. For generations, astronomers have been observing and gaining understanding the life cycle of stars. The evidence from observation is that stars have a life cycle very similar to humans: they are born, they live through distinguishable stages of life, they flame out, and they die, leaving behind the shells of their existence. It is in this process that objectivists believe the elements found in the Earth and other planets and asteroids are created. If in fact the elements that constitute all matter are created in the stars, it is easy to understand that a literal, twenty-four-hour day interpretation of the Genesis narrative seems unreasonable.

The star story is quite amazing: from the simplest element, hydrogen, all of the ninety-eight naturally occurring elements are assembled. An understanding of the life cycle of stars helps us understand that the elements, up to the mass of iron, are made in stars. Elements with greater mass than iron are formed in the supernova stage, which also distribute these elements throughout the universe. The remaining, human-made elements are assembled from the naturally occurring elements, which are commonly arranged in this illustrative chart.

For the sake of simplicity, we might compare these elements to a Lego set, which is comprised of similar pieces of different sizes, shapes, and colors. From these Legos, almost anything imaginable can be built.

According to most physicists, spectrograph evidence verifies that stars and supernovas are forming elements right now, so it becomes evident that the building blocks of the universe (or God's Lego set) is produced in the stars. It seems that an abundant amount of hydrogen (H) was created at the onset of the Big Bang. We must also assume that the four governing forces came into existence at that same moment. As the hydrogen spread throughout the universe, the force of gravity caused the hydrogen to collect together in massive nebula clouds. As a cloud develops, rotation follows, and the hydrogen molecules collect closer and closer together. The friction between molecules generates heat. Over time, the hydrogen reaches critical mass, and a new star is born.

In stars that generate temperatures greater than fifteen million degrees Fahrenheit, a nuclear chain reaction begins, and the process of nuclear fusion fuses two hydrogen atoms into one helium (He) molecule. In this process, a tremendous amount of energy is released. This is the process that was harnessed to produce fusion nuclear weapons and nuclear power plants. In high-mass stars reaching the red giant phase, where enough heat is available, the process continues, and carbon (three fused helium nuclei), oxygen (four fused helium nuclei), and iron (five fused helium nuclei) are also formed.

Once again gravity enters the picture, and the iron concentrates in the core of the star, causing fusion to slow and eventually cease. A battle between the cold, contracting mass of iron and the superheated, expanding gasses ensues. When temperatures approach ten billion degrees Fahrenheit, the star collapses, resulting in an incredible explosion called a supernova. In the heat of that explosion, physicists say the elements heavier than iron are formed.

We might call this the physicist's version of material evolutionary, thus giving rise to the old Earth theory, which is compatible with Darwinian evolution. Certainly, their evidence seems credible and verifiable and lends itself to a reasoned explanation of how the universe formed. Even though no one could observe the formation of the elements in the beginning, we can observe that process in real time right now, relegating it to science and not speculation. An understanding of nuclear fusion forms a firm scientific foundation for understanding how the stars are element factories. However, it leads creationists and evolutionists back to the instant before time began to the unanswered question. From where did time, space, matter, energy, and the four fundamental governing forces come? Objectivists say it was inevitability, IDC says it was intelligence, and theists say it was the Creator.

Once again, it is not the science but the interpretation of the event that makes Professor Hawking's *The Theory of Everything* become more viewpoint or philosophy than science. Let's be clear: evolutionists and creationists today agree that the universe has not always existed and that in an instant, space, time, matter, energy. and the four governing forces came into existence. It seems that Dr. Stephen Hawking became an unwilling participant in bringing creationists and evolutionists together in his refinement of the singularity principle.

Image Credit: Wolfgang Brandner (JPL/IPAC), Eva K. Grebel (Univ. Washington), You-Hua Chu (Univ. Illinois Urbana-Champaign), and NASA/ESA

NASA

Traditional creationists do not get hung up on specific timing issues. They do not hold to a literal, six-day creation period because they have created various theological gap theories to span enough time to be able to accept that the universe could be billions of years old. Old Earth creationists approach the creation accounts of Genesis in a number of different ways. The framework interpretation or framework hypothesis notes that there is a pattern, or framework, present in the Genesis account and that because of this, the account may not have been intended as a strict chronological record of creation. Instead, the creative events may be presented in a topical order. This view is broad enough that proponents of other old Earth views, including many day-age creationists, have no problem with many of the key points put forward by the hypothesis, though they might believe that there *is* a certain degree of chronology present.

Here's a summary of the Genesis six-day creation account, showing the pattern according to the framework hypothesis (from http://www.thefullwiki.org/Answers_in_Creation).

Day-age creationism is an effort to reconcile the literal Genesis account of creation with modern scientific theories on the age of the universe, the Earth,

Days of Creation	Days of Creation
Day 1: Light; day and night	Day 4: Sun, moon, and stars
Day 2: Sea and heavens	Day 5: Sea creatures; birds
Day 3: Land and vegetation	Day 6: Land creatures; man

life, and humans. It holds that the six days referred to in the Genesis account of creation are not ordinary twenty-four-hour days but rather are much longer periods of time.

Gap theories are quite another thing. Gap theologians create vast holes in the history of the universe to span immense periods of time. The most popular gap theory is that there is a gap of indeterminate time between

Genesis 1:1 and 1:2. One of the most developed theories proposed by the Churches of God Worldwide goes like this.

> That there was a pre-Adamic creation seems undeniable. The assertion, however, that this is contrary to the biblical narration does not stand up to scrutiny. One of the most fundamental errors of the theologians was that of Genesis chapter 1, verses 1–2. The term, "was waste and void" does not denote a condition of creation, but rather a condition of destruction. The earth became tohuw and bohuw, or waste and void; it was not created that way (cf. Isaiah 45:18). Simply put, the earth has existed for many millions of years. The Genesis story is a re-creation story, not the total creation story. The gap between Genesis 1:1 and Genesis 1:2 is untold millions of years.

This is but one of hundreds of theological explanations for gaps of time that permit traditional creationists to accept an old Earth. But as was previously noted, new Earth creationists believe that the Earth could not possibly be more than six to ten thousand years old. Don't be surprised, but their model is also based in science. Russell Humphries has been in the laboratory and discovered his own body of evidence, as summarized below.

Too many tightly grouped galaxies

The stars rotate about around the galaxy's center at different velocities, the inner rotating faster than the outer. If galaxies were more than a few hundred million years old, they would become a disorganized disc rather than a tight spiral. Evolutionists have tried to explain the discrepancy, but the Hubble Space Telescope's discovery of a very tight spiral structure in the central hub of the Whirlpool Galaxy is indisputable proof that the universe is relatively young.

Too few supernova remnants and too many comets

Astronomers observe approximately one supernova every twenty-five years. The remnants from such explosions (like the Crab Nebula) expand outward rapidly and theoretically remain visible for a million years. Yet we can observe only about two hundred supernova remnants. You do the math! The same is true with comets, which lose much of their mass every time they pass the sun. If the comets were formed at the same time as the universe, as evolutionists say, none should still survive.

Too little mud and too much sodium in the oceans

Each year, through erosion, about twenty billion tons of dirt and rock get deposited on the ocean bottom. Tectonic plate shift removes about 5 percent of this material. The depth of sediment in the whole ocean averages less than four hundred meters, which represents about six thousand years of debris. Consequently, rivers and streams dump five hundred million tons of sodium into the oceans each year. Even though 25 percent is removed, the oceans are not nearly as salty as they should be if the oceans are three billion years old, as evolutionists contend.

Too much energy remains in the Earth's magnetic field

Another simple mathematics issue is the amount of remaining energy in the Earth's magnetic field, which decreases every year. Although evolutionists have devised a crafty answer based on faulty physics, at the current rate of decay, the Earth could not be more than nineteen thousand years old.

Too many radically bent strata layers

In mountainous regions, strata layers thousands of feet thick are bent and folded into hairpin shapes, without cracking, meaning the entire formation had to be wet and unsolidified when the bending occurred. This means that the folding occurred thousands, not millions, of years after being deposited.

Too much intact DNA in specimens thought to be millions of years old

DNA and other biological material rapidly decompose. Measurements of the mutation rate of mitochondrial DNA recently forced researchers to revise the age of "mitochondrial Eve" from a theorized two hundred thousand *years* down to possibly as low as six thousand years. DNA experts insist that DNA cannot exist in natural environments longer than ten thousand years, yet intact strands of DNA appear to have been recovered from fossils that are allegedly much older.

Too much carbon-14 and too little helium in deep geologic strata

Carbon-14 atoms, which have a short life, should not exist for more than 250,000 years. Yet Ice Age strata contain significant amounts of carbon-14. Uranium and thorium generate helium atoms as they decay to lead. A study published in the *Journal of Geophysical Research* showed that such helium produced in zircon crystals in deep, hot, Precambrian granitic rock has not had time to escape. Newly measured rates of helium loss from zircon show that the helium has been leaking for only six to eight thousand years. This is very strong evidence that the earth is only thousands, not billions, of years old.

Too few skeletal remains

Evolutionists contend that *Homo sapiens* have roamed the Earth for two hundred thousand years and that population estimates for that period were consistent, between one and ten million. Those calculations suggest that at least eight billion bodies were buried or burned, often with artifacts. Only a few thousand have been found, which implies that the Stone Age was much shorter than evolutionists think, possibly less than a thousand years.

Too long for agriculture, written language and monumental architecture to develop

There is evidence that agriculture was developed less than ten thousand years ago and that written language and monumental architecture was developed five thousand years ago. Why did it take 185,000 years for those basic components of survival to be discovered? Prehistoric man built megalithic monuments, made beautiful cave paintings, and kept records of lunar phases. Isn't it logical that of the eight billion people mentioned in the previous point, which evolutionists contend were intelligent, some would have discovered that plants grow from seeds, buildings provide protection, and language must be written?

Dr. Humphries' information presents some challenges to an old universe, so who is right? Is the earth 6,000 years old or 13.8 billion years old? What if the answer to that question is, "Yes"? Australian physicist Dr. Barry Setterfield may be able to support such a notion. Einstein suggested that time is relative, but Setterfield takes this idea to the extreme by generating a convincing set of arguments and a theory that suggests the orbital age of the Earth (the time it takes the Earth to make a complete orbit of the sun) vastly differs from its atomic age, which is measured by the rate of atomic delay. Without controversy, science agrees that every accurate measurement concerning the atomic age of the universe and the matter in it depends upon the speed of light being constant. Dr. Setterfield's theory is that the constant is not constant at all—the speed of light is actually slowing down and has done so since the beginning of the universe, in exponential proportions. Thus, the atomic age of the earth is not even close to its orbital age. In Setterfield's theory, billions of years of atomic time could fit nicely into thousands of orbital years. If the constant is not really constant, it is a new ballgame for time calculation, and that impacts the geological record and destroys the carbon-dating scale.

The basis of his theory is based in two thoroughly explored astronomical principles, variations in both the red shift and Planck's constant. He claims that at some point in time, the Earth stopped expanding—an idea rejected by the mainstream scientific theory, which proposes that the universe has expanded for billions of years and continues to expand. The problem that Setterfield's theory creates is that most scientists use the rate of radioactive decay of certain elements as a scale for determining elapsed time. All ordinary matter is made up of combinations of chemical elements, each with its own atomic number, indicating the number of protons in the atomic nucleus. Additionally, elements may exist in different isotopes, with each isotope of an element differing in the number of neutrons in the nucleus. A particular isotope of a particular element is called a nuclide. Some nuclides are inherently unstable. That is, at some point in time, an atom of such a nuclide will spontaneously transform into a different nuclide. This transformation may be accomplished in a number of different ways, including radioactive decay, either by emission of particles or by spontaneous fission and electron capture.

Although the moment in time at which a particular nucleus decays is unpredictable, a collection of atoms of a radioactive nuclide decays exponentially at a rate described by a parameter known as the half-life, usually given in units of years when discussing dating techniques. After one half-life has elapsed, one-half of the atoms of the nuclide in question will have decayed into a "daughter" nuclide or decay product. In many cases, the daughter nuclide itself is radioactive, resulting in a decay chain, eventually ending with the formation of a stable, nonradioactive daughter nuclide; each step in such a chain is characterized by a distinct half-life. In these cases, usually the half-life of interest in radiometric dating is the longest one in the chain, which is the rate-limiting factor in the ultimate transformation of the radioactive nuclide into its stable daughter. Isotopic systems that have been exploited for radiometric dating have half-lives ranging from about ten years for tritium to over one hundred billion years for samarium.

For most radioactive nuclides, the half-life depends solely on nuclear properties and is essentially a constant. It is not affected by external factors such as temperature, pressure, chemical environment, or the presence of a magnetic or electric field. This predictability allows the relative abundances of related nuclides to be used as a clock to measure the time from the incorporation of the original nuclides into a material to the present. But if the rate of decay, which is calculated directly from the speed of light, has declined exponentially through the ages, the scale is fatally flawed. The speed of light is a sacred cow in science and is the benchmark upon which most of the great theorems are based. It is the basis for Einstein's major discovery, $E = mc^2$.

Below is a more detailed description of Dr. Setterfield's work.

> The speed of light has been measured 163 times by 16 different methods over the past 300 years. However, Australian physicist Barry Setterfield and mathematician Trevor Norman have suggested the speed of light appears to have been slowing down!
>
> 1657:   Roemer   307,600*

| 1875: | Harvard | 299,921 |
| 1983: | NBS | 299,792.4358 |

*km/sec

Alan Montgomery's computer analysis supports the Setterfield/ Norman results and indicates that the decay of velocity closely follows a cosecant-squared curve and has been asymptotic since 1958. If correct, the speed of light was 10–30% faster in the time of Christ; twice as fast in the days of Solomon; four times as fast in the days of Abraham, and perhaps more than 10 million times faster prior to 3000 BC. (http:// www.khouse.org/articles/1999/225).

A graph from Setterfield's own website illustrates that the speed of light dropped very swiftly at first, then more slowly as one approaches the common age. This is a deal changer for creationists, especially young Earth creationists. Ken Ham is a staunch supporter of Setterfield's research and conclusions. Although this theory is extremely complex and requires some degree of expertise in physics, Setterfield's lectures on YouTube give an exhaustive presentation of the theory and seem rational and reasonable. The fact that the relationship between the atomic year and the solar year have never been explored speaks to a certain closed-mindedness often displayed by the modern scientific community. Other Setterfield assertions, that the universe is no longer expanding, and that Planck's constant is slowing, also deserve more research and study.

So far, we have looked at the age of the Earth through the lens of the physical sciences, yet there is biological, anthropological evidence that must be considered. Is there clear evidence of a time when intelligent, rational humans like us appeared on the Earth? Once again, evolutionists say one hundred thousand years, and six-day creationists say six thousand years. But what do the facts say?

A recent scientific study of human genetics related to the brain concluded that a unique ASPM (Asporin) variant appeared 5,000–15,000 years ago. This roughly correlates with the rapid development of written language, monumental architecture, refined agriculture, and the consequential rise of civilization and cities. This lends support to the "Fortuitous Mutation" that Richard G. Klein refers to in his book *The Dawn of Human Culture,* in which he says this mutation occurred about 50,000 years ago.

Before six-day creationists throw this theory in the trashcan, forget the timing for now and remember that all we are saying here is that there is the possibility of the rapid improvement of humankind's ability to adapt to a more complex civilization. This could explain the explosion of technology in the days when Nimrod built the first great city, Babel. Furthermore, there is more modern evidence that suggests that microevolution or adaptation continues, especially as it relates to the human brain.

> Human evolution—in what has become our most important organ, the brain—is still under way. Two studies show genes linked to brain size are rapidly evolving in humans. "Our studies indicate that the growth of brain size and complexity—is likely still going on," said lead researcher for both papers, Bruce Lahn, an investigator in the Howard Hughes Medical Institute. "Meanwhile, the skills we need to survive in it are changing. I would expect the human brain, which has done well by us so far, would continue to adapt.[1]

All of this information fits neatly into universally accepted views of microevolution, or adaptation within individual species. It seems that the human skills necessary for adapting to civilization are continually improving. For creationists, these specific facts lead to the conclusion that thoroughly modern humans arrived in the form of Adam and Eve, who were created in God's image, with the ability to adapt to their environment. Evolutionists might say that this was the genetic breakthrough that differentiated Neanderthal or Denisovan man from modern man.

---

[1] Catherine Gianaro, "Lahn's Analysis of Genes Indicates Human Brain Continues to Evolve." More information on Dr. Lahn's research is available from his University of Chicago website: https://genes.uchicago.edu/directory/bruce-lahn-phd.

Before we proceed, let us take a look at comments from respected historians and archeologists regarding the contributions of the Bible to the study of ancient history.

Smithsonian Anthropology Department

The historical books of the Old Testament, are as accurate historical documents as any that we have from antiquity and are in fact more accurate than the Egyptian, Mesopotamian, or Greek histories. These Biblical records can be and are used as are other ancient documents in archeological work since historical events described took place and the peoples cited really existed.

National Geographic Society

I referred your inquiries to our staff archeologist, Dr. George Stuart. He said that archaeologists do indeed find the Bible a valuable historical reference tool, and use it many times for geographical relationships, old names and relative chronologies.

R. D. Wilson

My recent book, *A Scientific Investigation of the Old Testament* pointed out that the names of 29 Kings from ten nations, including Egypt, Assyria and Babylon are mentioned not only in the Bible but are also found on monuments of their own time. Every single name is transliterated in the Old Testament exactly as it appears on the archaeological artifact—syllable for syllable, consonant for consonant. The chronological order of the kings is also correct.

Sir William Ramsay

Ramsay, an outstanding Near Eastern archeologists, says: "Luke is a historian of the first rank; not merely are his statements of fact trustworthy; he is possessed of the true historic sense; he fixes his mind on the idea and plan that rules in the evolution of history, and proportions the scale of his treatment to the importance of each incident. He seizes the important and critical events and shows their true nature at greater length ... In short, this author should be placed among the very greatest of historians."

Jack Wellman

Archaeological support for the Bible continues to swell, year after year. Today, well over 50,000 digs at 30,000 different locations have been excavated. These finds continue to provide substantial support of the Bible's written record. Some of the finds include Belshazzar, King Darius, and King Cyrus. What were once thought to be mythological empires mentioned in the Old Testament, like the Hittite's, have been found to have existed after all.

Because many unbiased scholars accept the Bible as an ancient historical source that has proven to be helpful to historians and archeologists, let's examine the biblical record of the development of agriculture, previously mentioned in chapter 1, which was the cornerstone of the rise of the most ancient civilizations (Sumer, Egypt, and the Indus Valley Culture). The Bible repeatedly says that Adam and his offspring were farmers.

And the Lord God took the man, and put him into the garden of Eden to dress it and too keep it. (Genesis 2:15)

Therefore the Lord God sent him [Adam] forth from the garden of Eden to till the ground from whence he was taken. (Genesis 3:23)

And Abel was a keeper of sheep, but Cain was a tiller of the ground. (Genesis 4:2)

Can a case be made from archeological data that validates the claims of creationists?

Biblical Research 🗿 🗿 *Biblical Archaeology*

The Evidences for a Recent Dating for Adam, about 14,000 to 15,000 years Before Present

The great majority of the cultivated plants of the world trace their origin to Asia. Out of 640 important cultivated plants, about 500 originated in Southern Asia. Studies demonstrate that Asia is not only the home of the majority of modern cultivated plants, but also of our chief domesticated animals such as the cow, the yak, the buffalo, sheep, goat, horse, and pig.

More recent studies conducted by Melinda A. Zeder place the initial domestication of goats to the Zargos Mountains at about 10,000 years ago, pigs at 10,500 and cattle at 10,000. These dates mean that animals were domesticated at much the same time as crop plants, and bear on the issue of how this ensemble of new agricultural species—the farming package known as the Neolithic revolution—spread from the Near East to Europe. Studies indicate that large scale wheat cultivation began from 8,000 to 9,000 years ago in Mesopotamia.

Simcha Lev-Yadun, of Israel's Agricultural Research Organization, and colleagues suggest the first farms may have been between the Tigris and Euphrates rivers in what is today northeastern Turkey and northern Syria. Archaeological evidence shows that the earliest known farming settlements of the Fertile Crescent were in this core area. Also, the limited genetic variability of these crops implies that they were domesticated only once rather than by several different cultures at roughly the same time, about 9,300 years ago.

Genesis 11:2 And it came to pass, as they journeyed from the east, that they found a plain in the land of Shinar, and they dwelt there.

Archaeological evidence shows that around this time, farming techniques began to spread out of Anatolia, where the Ark is thought to have landed, across Europe and Asia. As already shown, the data on the farming indicates that the after the flood Genesis history took place in the Ararat area and that the area is also the origin of many of the known farm crops and domesticated animals.

As is always the case, climate variation had a dramatic effect on the spread and advancement of agriculture in Europe. Since 1916, with the onset and refinement of pollen analytical techniques, palynologists have concluded that the Younger Dryas represented a distinct period of vegetational change in large parts of Europe during which warmer climate vegetation was replaced by generally cold climate glacial plant succession.

So, what would greatly increase the toil of a group of farmers more than a period of severe climate? So, it would seem that one could conjecture that the period of a relatively warm period of about 14,000 years ago was when Adam started farming. This warming trend was followed by a cool period of from 14,000 years ago to about 12,000, which could correspond to "the curse of the ground" spoken of in Genesis, chapter 3, a period in which farming was more difficult. Then about 12,000 years ago the warming begins again, and farming becomes easier and proliferates. The point is that six-day creationists do not rest their case in theology alone; their views have some basis in genuine academic thought as is shown in this comparison of the Biblical record with agricultural development.

Unfortunately, understanding "the beginning," does not get much help from the original Old Testament language, Hebrew. The Hebrew word *yom* usually translated as day, can refer to a twenty-four-hour day (sunrise to sunset) or a long, unspecified period of time as in "the day of the dinosaurs." In addition, the Hebrew word *ereb*, translated as evening, also means sunset, night, or ending of the day. The Hebrew word *boqer*, translated as morning, also means sunrise, "coming of light," "beginning of the day," or dawning. Our English expression "The dawning of an age" serves to illustrate this point. Likewise, this expression in Hebrew could use the word *boqer* for dawning, which in Genesis 1 is often translated as *morning*. The actual number of words in Hebrew is much fewer than that of the English translations. The words "and there was" are not included but are added to make the English flow better. The actual translation is "evening and morning 'n' day." There is no way to discern from the verbal context that the text is referring to twenty-four-hour days.

The first thing one notices when looking at Genesis 1 is the unusual construction surrounding the words *morning* and *evening* together with *day*. This combination is very rare, occurring only ten times in the Old Testament, six of which are in the Genesis creation account. Not all instances of *morning* and *evening* refer to a literal period of time. Here are two specific examples.

1. In the morning it [grass] flourishes, and sprouts anew; toward evening it fades, and withers away. (Psalm 90:6)

This verse refers to the life cycle of grass compared to the short life span of humans. Obviously, the grass does not grow up in one morning and die by the same evening. The period of time refers to its birth in the morning and its death in the evening, at least several weeks or months later.

2. And the vision of the evenings and mornings which has been told is true; but keep the vision secret, for it pertains to many days in the future. (Daniel 8:26)

This verse actually refers to events that are yet to happen, 2,500 years from the time when it was originally written. One could easily say that these mornings and evenings represent thousands of years. The apostle Peter tells us, "With God, a thousand years is as a day" (2 Peter 3:8). From a logical perspective, the third day of creation must have been longer than twenty-four hours. The text indicates a process that would take years. On this day, God allowed the land to produce vegetation and trees, which bore fruit. The text specifically states that the land produced trees that bore fruit with seeds in them. Any horticulturist knows that fruit-bearing trees require several years to grow to the maturity to be able to produce fruit. However, the text states that the land produced these trees (indicating a natural process) and that it all occurred on the third day. Obviously, such a "day" could not have been only twenty-four hours long.

From a purely rational standpoint, the events of the sixth day of creation, which is the only day for which we have exhaustive detail, seem to require more than a literal, twenty-four-hour day. On this day, God created the mammals and humankind. The details of His careful fashioning of Adam from earthen clay indicates great care. Then He placed Adam in his new home, an expansive garden; gave him a tour; thoroughly instructed him on its care; and finally gave him specific warnings about the Tree of the Knowledge of Good and Evil. Next, God brought all the species of animals to Adam to be named. This job in itself could have taken many days, weeks, or months. After a suitable companion for Adam could not be found among the animals, God put Adam to sleep, took out a rib, and created Eve. When Adam woke up, he used the Hebrew word *pa'ămâh*, which means "at long last." If Adam had only waited a few hours for his helpmate, it is unlikely he would have used this Hebrew word. The context suggests that Adam had to endure loneliness before Eve was created, so it is very unlikely all of this could have taken place in twenty-four hours, especially because much of it was dependent upon Adam, who did not have the same abilities as God.

The belief that creation days are long periods of time is not a recent interpretation of the scriptures; it has been prevalent since the first century. Dr. Hugh Ross published a book in 1994, entitled *Creation and Time*, that documents in detail what first-century Jewish scholars and the early Christian Church fathers said regarding their interpretation of creation chronology. Jewish scholars include Philo and Josephus, and Christian fathers include Justin Martyr, Irenaeus, Hippolytus, Clement, Origen, Lactantius, Victorinus, Methodius, Augustine, Eusebius, Basil, and Ambrose. Among this group, nearly all acknowledged the likelihood that the creation days were longer than twenty-four hours. The evidence presented in *Creation and Time* is both compelling and well documented. I highly suggest you read it.

Thorough investigation of this evidence, we conclude that "the beginning" holds no answers to our inquiry. One way or another, it does seem clear from the evidence that time, space, energy, matter, and the four fundamental forces came into being suddenly and completely, but the when remains a question. Because we all agree that matter changes but is neither created nor destroyed, the Big Bang seems to be a satisfactory name for what happened in the beginning. The unanswered question is whether the Big Bang was an intentional act of creative intelligence or is an unexplainable cosmic accident that mysteriously and unexplainably occurred 13.4–13.7 billion light-years ago.

CHAPTER 3

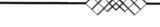

SEEKING ANSWERS FROM LIFE

If you think that the age of the universe is controversial, what about the age and origin of life? Professor Stephen Hawking, one of the world's foremost cosmologists, put it this way in his documentary *Into the Universe: The Story of Everything* (2010): "I am amazed at how much we understand about the origin of the universe, which occurred billions of years ago … and how little we know about the origin of life, which is far more recent." He goes on to say that life remains a mystery, but our best guess is that it was simply a cosmic accident. Is this not an admission by the most brilliant mind of the twenty-first century that the question of the origin of life is still unanswered?

Another brilliant scientist and a vocal leader of the Darwinian evolutionary viewpoint, Richard Dawkins, said something similar in an interview recorded for the movie *Expelled*. In the absence of any cause or observable process for the creation of life, he admitted the possibility that life could have been "seeded" on Earth by aliens, as is suggested by the Hollywood film *Prometheus*. This is not what he really believes, but in the absence of clear evidence of spontaneous generation, he would rather admit that alien seeding is more possible than suggesting that it might have been a supernatural creative act.

So why the confusion? All living creatures have some sort of DNA and are otherwise similar. Functions like respiration and reproduction are shared by all life forms. Yet there are vivid contrasts, uniqueness, and unexplained complexities in even the simplest life. This is no truer than with humankind. It is not just intelligence, language, opposable thumbs, conscience, creativity, self-awareness, and morality that make us unique; there is the intangibleness of conscious awareness. From the days of Plato and Aristotle, insightful men have been analyzing the difference between us and the animals, and there is universal acceptance that we are a distinct and special species.

Creationists and evolutionists would all agree that species adaptation, also called microevolution, natural selection, or "survival of the fittest," is good science. Microevolution, the small, gradual changes within a species that are beneficial for survival, cannot be denied. This phenomenon is scientifically observable at the physical and genetic levels. Darwin is to be commended for his work in analyzing and defining the process of microevolution. That's pretty good for a theologian.

There is evidence that the human race has adapted through the ages. There is no denying that we are vastly different from our ancestors and even from other humans living in different parts of the world. Through the millennia, microevolution has helped us survive as a species. If conditions demanded it, we may have been more stooped, more muscular, more agile, more brutish, or more dexterous, but we remain *Homo sapiens*. The Zulus and the Pygmies, who live hundreds of miles apart in Central Africa, represent the extremes of human variety. The Bushmen are simple, nomadic, hunter-gatherers living mostly in the forest. The Massai, their neighbors, are mighty hunters capable of incredible physical feats; their physical size, strength, cunning, and athletic abilities clearly demonstrate adaptation.

Dr. Orville Boyd Jenkins writes,

> History indicates all humans in every place have a rich mix of genetic and cultural heritage. People and peoples are as we find them. We can only begin to probe that in various ways. There is a wide range of physical types among the peoples of Eastern Africa … the truth that the primary thing the various peoples of the Horn of Africa have in common is that they all live in Eastern Africa! This generally means that the population groups that now live in the African continent exhibit great diversity from one to the other. The focus is not on the variation within any one population group, but the variation among different population groups.

A 2012 article describes the complete genomic sequence of three African hunter-gatherer populations, the Hadza and the Sandawe from Tanzania and the Pygmies from Cameroon. The researchers were looking for telltale variants in the genetic code that could help explain differences between individuals and populations. Among them were sharp differences between the three groups in smell and taste, suggesting that "each population's senses had adapted to the new smells and foods they encountered."

The natural genetic variation within a population of organisms means that some individuals will survive more successfully than others within their environment. Factors that affect reproductive success are also important, an issue which Charles Darwin developed in his ideas on sexual selection. Natural selection acts on the phenotype, or the observable characteristics of an organism, but the genetic basis of any phenotype that gives a reproductive advantage will become more common in a specific population. Over time, this process can result in adaptations that specialize organisms for particular ecological niches: heat, cold, wet, dry, rich food source, or poor food source. Evolutionists point out the size of the human skull as one important adaptation. An interesting fact about the skulls of the supposed predecessors of humans is that those discovered were as large, or in some cases larger, than those of modern humans.

This largely unexplained anomaly is an issue you don't hear a lot about, at least not in the mainstream, but the answer to this puzzle might be vital for the equity of our journey, so let's open another Pandora's box.

In 1993, Michael Cremo, a George Washington University graduate and former naval officer, cowrote *Forbidden Archeology* with Richard L. Thompson. This book makes the point that the modern human is not modern at all. In fact, it claims that humans like us have lived on the earth for millions of years and that the scientific establishment has suppressed the fossil evidence of extreme human antiquity. He speaks about a knowledge filter as the cause of this suppression. Because Cremo is a Hindu, the book has attracted attention from some mainstream scholars as well as Hindu creationists. Although some scholars of mainstream archeology and paleoanthropology have described the work as pseudoscience, some credible critics have acknowledged positive aspects of the book. Anthropologist Kenneth L. Feder wrote in his review,

> While decidedly antievolutionary in perspective, this work is not the ordinary variety of anti-evolutionism in form, content, or style. In distinction to the usual brand of such writing, the authors use original sources and the book is well written. Further, the overall tone of the work is far superior to that exhibited in ordinary creationist literature.

Archaeologist Tim Murray also wrote a review of *Forbidden Archeology*.

> "I have no doubt that there will be some who will read this book and profit from it. Certainly it provides the historian of archeology with a useful compendium of case studies in the history and sociology of scientific knowledge, which can be used to foster debate within archaeology about how to describe the epistemology of one's discipline." He also commented on the similarities in argument with those of Christian Creationists: "This is a piece of 'Creation Science' which, while not based on the need to promote a Christian alternative, manifests many of the same types of argument: first, an attempt to characterize the opposition as motivated by the need to preserve their view of the world rather than a desire to practice unfettered inquiry; secondly, to explain

the currently marginal position of your alternative as being the result of prejudice, conspiracy and manipulation rather than of any fault of the theory itself; thirdly, to present the opposition, in this case mainstream palaeoanthropology and quarternary archaeology, as being united as a 'secret college' to manipulate the public mind and to exclude non-professionals from being able to control science for the benefit of all."

Cremo identifies himself as a "Vedic archeologist" because he believes his findings support the story of humanity described in the Vedas. The Indian magazine *Frontline* called Cremo and Thompson "the intellectual force driving Vedic creationism in America." Historians of science Jo Wodak and David Oldroyd published a twenty-three-page review article, "Vedic Creationism: A Further Twist to the Evolution Debate in Social Studies of Science." The review points out positive aspects of the book: that *Forbidden Archeology* brings to attention many interesting facts that have not received much consideration from historians, and that the authors' detailed examination of the early literature is stimulating and raises questions of considerable interest, both historically and from the perspective of practitioners of sociology of scientific knowledge. Cremo's books provided much of the content for the widely criticized 1996 NBC special *The Mysterious Origins of Man*.

Like Setterfield's model of time, Cremo's archeological evidence could be a game changer in our quest for truth. It barges through strongly guarded doors and demands us to consider the implications of such a world. This gives new dimension to micro evolution and how humankind has become what it is today. It also might answer other unexplained questions from a mysterious time before antiquity that produced marvels like the Stones of Puma Punku, the Nazca Lines, the Piri Reis Map, and the Antikythera Computer, but that is material for another book.

Yet stretching Darwin's theory to the macro scale of species migration is unacceptable for creationists. Phillip Johnson goes to the heart of the matter.

Critics of evolutionary theory are well aware of the standard examples of microevolution, including dog breeding and the cyclical variations that have been seen in things like finch beaks and moth populations. The difference is that we interpret these observations as examples of the capacity of dogs and finches to vary within limits, not of a process capable of creating dogs and finches, much less the main groups of plants and animals, in the first place … "As any creationist (and many evolutionists) would see the matter, making the case for 'evolution' as a general theory of life's history requires a lot more than merely citing examples of small-scale variation. It requires showing how extremely complex biological structures can be built up from simple beginnings by natural processes, without the need for input or guidance from a supernatural Creator." (Reason in the Balance, 1995, p. 74)

Creationists do not believe Darwin's assertion that microevolution is followed by macroevolution, the change of one separately distinct species to another separately distinct species. In other words, creationists believe that a wolf, *Canis lupus*, can become a bulldog, *Canis lupus familiaris*, but not a pig, *Sus scrofa domesticus*. A dog and a wolf are the same species because they can successfully breed and bear young that are able to reproduce, whereas dogs and pigs are two separately distinct species and cannot breed. Mules and horses are separate species and can breed and produce offspring, but that offspring is incapable of reproduction.

Evolutionists, both objectivist and theistic, believe that the entire process of macroevolution is observable science. Francis Collins, the man who unlocked the human genome, declares, "Evolution is no more a theory than that of a round earth." What confidence! The science of evolution does have scientific challenges, however. Creationists say that it may not be as cut-and-dried as Dr. Collins supposes. DNA may be God's Language, but creationists might say that this language was written one species at a time. The detailed description of every living species is written with just four simple phosphate molecules, typically described as AGTC. That fact alone is used by the creationist as teleological proof for the existence of a creator and is used by objectivists to support the theory of evolution.

Could an animal become thoroughly human over a vast evolutionary span? Could humans develop all of the characteristics necessary for survival through a developmental evolutionary process? Much has been written on the subject. In 1900, De Vries, Tschermack, and Correns used Mendel's work to support Darwin's assertion that it could. Dennis Bonnette, the resident theologian for theistic evolution, writes, "Evidence favoring the interconnectedness of species seems overwhelming ... biochemical evidence, when mentioned with other information (the evolutionary path of Eohippus) has been so complete that we tend to dismiss any opposition to evolution."

That being said, evolutionists bear the burden of proof. The theistic evolutionist faces the same challenge, as does the objective evolutionist. Simply believing in a creator does not get one off the hook. Uncontroverted, empirical evidence for evolution must be demonstrated because the similarity of genetic code could be explained by each species having the same designer. A Ford is recognizable as a Ford, and a Rolex is recognizable as a Rolex. There are distinguishing characteristics of each specific model, but there is also evidence of similar design and manufacturing. Along those lines, much has been said and written about the genetic similarity of chimpanzees and humans. Up until recently, we were told by geneticists and anthropologists that the two species are 98.5 percent similar, but once again, new exculpatory evidence has arisen.

I have written about the similarity between human and chimpanzee DNA three times before. It's an important question for creationists, intelligent design advocates, and evolutionists alike, since the chimpanzee is supposed to be the closest living relative to human beings. As a result, a comparison of chimp DNA to human DNA gives us some idea of what the process of evolution would have to accomplish to turn a single apelike ancestor into two remarkably different species like chimpanzees and people.

Early on, it was widely thought that human DNA and chimp DNA were 99% similar. As I discussed in my first post on this subject, that was based on a very limited analysis of only a minute fraction of human and chimp DNA. Now that the entire set of nuclear DNA (collectively called the "genome") of both humans and chimpanzees have been sequenced, we now know that the 99% number is just plain wrong. Interestingly enough, however, even though both genomes have been fully sequenced with a reasonable amount of accuracy, no one can agree on exactly how similar the two genomes are.

Why is that? Because comparing genomes is a lot harder than you might think. While we know the sequence of the chimp and human genomes really well, we don't understand the DNA itself. Indeed, there are large sections of DNA that seem to be functional, but we simply have no idea what they do. As a result, comparing the genomes of two different species can be very, very tricky.

Probably one of the best explanations of just how tricky DNA comparison is comes from Dr. Richard Buggs, a geneticist at Queen Mary, University of London. Back in 2008, he wrote about the steps he would take to compare the human and chimp genomes, and if you read his explanation, you will get an idea of how difficult such a comparison is. His conclusion was:

Therefore the total similarity of the genomes could be below 70%.

Since that time, the chimpanzee genome has been sequenced to an even better degree, and other methods have been used to determine the similarity between the chimpanzee and human genomes. One of the more popular methods is based on an algorithm called BLAST, which chops up DNA (or proteins) into small segments and then tries to compare them to the segments on a different set of DNA (or proteins). This seems like the most "generous" way to compare two genomes, because it doesn't require one genome to be structured similarly to the other. The only thing that matters is whether a bit of information in one genome can be found anywhere in the other genome.

Using this method to determine the similarity between the human and chimp genome, researchers have come up with different answers. Dr. Todd Wood, an expert in genome comparison and former Director of

Bioinformatics at the Clemson University Genomics Institute, did a BLAST analysis that indicated human and chimp DNA are roughly 95% similar. However, Dr. Jeffrey P. Tomkins, former director of the Clemson University Genomics Institute, did a different BLAST analysis and concluded that the similarity was 86–89%.

Well, Dr. Tomkins just published a new study, and as far as I can tell, it makes the most sense of any BLAST analysis done so far. In this study, he chopped up the chimpanzee genome into "slices" that were as small as 100 base pairs long or as large as 650 base pairs long. The chimpanzee genome is 2.9–3.3 billion base pairs long, so obviously these slices are incredibly small compared to the entire genome. He then looked for each "slice" on the human chromosome that is supposed to correspond to the chimp chromosome where the slice was found. The two slices didn't have to match exactly; they just had to be similar enough to think that they could be related to each other.

The graph at the top of this post shows his results. Notice that the similarity hovers around 70% for all chromosomes except the Y chromosome. The size of the "slice" affects the result a bit, but really not much. In the end, this leads Dr. Tomkins to conclude:

"Genome-wide, only 70% of the chimpanzee DNA was similar to human under the most optimal sequence-slice conditions. While chimpanzees and humans share many localized protein-coding regions of high similarity, the overall extreme discontinuity between the two genomes defies evolutionary timescales and dogmatic presuppositions about a common ancestor."

Abstract

Using the new 6X chimpanzee assembly, a sequential comparison to the human genome was performed on an individual chromosome basis. The chimpanzee chromosomes, were sliced into new individual query files of varying string lengths and then queried against their human chromosome homolog using the BLASTN algorithm. Using this approach, queries could be optimized for each chromosome irrespective of gene/feature linear order. conservative because it did not include the amount of human DNA absent in chimp nor did it include chimp DNA that was not aligned to the human genome assembly (unanchored sequence contigs). For the chimp autosomes, the amount of optimally aligned DNA sequence provided similarities between 66 and 76%, depending on the chromosome. In general, the smaller and more gene-dense the chromosomes, the higher the DNA similarity—although there were several notable exceptions defying this trend. Only 69% of the chimpanzee X chromosome was similar to human and only 43% of the Y chromosome.

This study should remind us that creationists and IDT scientists have been asking for more physical evidence for macroevolution—the irrefutable proof, if you will. Although evolutionists have long said that the fossil record is evidence, creationists say it shows an explosion of living things (the Cambrian Explosion) ordered as separate, distinguishable species over a short period of time. They would say that this is the smoking gun that kills Darwin's theory of macroevolution and neutralizes evolutionists' claims.

It might surprise you to learn that many Christians are evolutionists. These theistic evolutionists accept the Genesis narrative as a sort of allegory to understand the evolutionary process for humankind, but choosing that position demands the same kind of evidence. Furthermore, it requires an explanation of how a dumb animal became human. Two popes (Pius XII and John Paul II) and one ecumenical council made room for Roman Catholics to accept theistic evolution. Unfortunately for those who go down this pathway, the popes and the council did not remove responsibility for "keeping the sacredness of the gospel intact," and therein lies the problem. For the Christian Gospel to work, numerous conditions relating to Adam, Eve, and Jesus Christ must be clearly established. Catholics, who genuinely believe the plan for the redemption of humankind and earnestly desire to uphold the important theological implications clearly revealed in scripture, must grapple with this.

Those conditions begin with Adam and the essential theological state of being made "in the image of God." Then the theology of redemption, or soteriology, continues with Eve coming directly from Adam ("bone of my bone and flesh of my flesh"). Following that, both Adam and Eve must have lived in the state of innocence and willfully sinned against God to receive the justly pronounced curse of a sin nature and the curse of death (both spiritual and physical). Lastly, every human who has ever lived must have come directly from them as descendants, especially the Redeemer Himself.

With each of these elements in place, Jesus Christ, the seed of the woman described in Genesis 3:16 as direct descendent of Adam (Son of Adam), could represent the entire human race on the cross of redemption just as Adam represented our entire race in the fall. As one of the first popes, the Apostle Paul, put it, "for as in Adam all die so in Christ shall all be made alive" (Romans 5:18). I know that for many of you this means nothing, but for Christians who decide to be evolutionists, it places a burden of immense consequence. How do you reconcile these specific and critical requirements with evolution? We will examine this more fully in the next chapter.

Finally, it must be stated that the mere proof of the existence of humanoids similar to humankind does not alone prove evolution. There are published theories that propose it's possible that there was a time when the earth was inhabited by three or more separate and distinct humanoid species that never shared an evolutionary link. David Pilbeam, in his article "Human Origins and Evolution" (Edited by A. C. Fabian, Cambridge, 1988, p. 111), talks of human development, commencing from *Homo erectus*, the first hominid species to achieve a wide distribution outside Africa.

> H. Erectus and its industries lasted from over 1.5 my ago to less than 0.5 my ago without much change, implying a degree of behavioral stability which was surprising when it first became apparent to paleoanthropologists. Around 0.3 my ago, hominids with somewhat larger brains (1100–1300 cm3) and slightly differently shaped skulls began to appear: so-called archaic Homo Sapiens; Overall, they were still very similar to Homo Erectus. To distinguish them from us, the

Modern Homo Sapiens, the archaic modifier is added. The best known archaic sapiens came from Europe and West Asia: the Neanderthals. However, despite their sapient label, despite attempts to "launder" them, I do not think they should be called Homo Sapiens because morphologically, and by inference, behaviorally they differ markedly from modern humans.

Pilbeam goes on to say (pp.112–113),

A substantial evolutionary change comes between archaic and modern Homo Sapiens. It occurred over the period between 130 ty and 30 ty ago, ending with the appearance of humans like us, with less massive skeletons, less muscular bodies, rounder heads and flatter faces. Judging from their archeological traces, their behavioral potential was also like ours by at least 35 ty ago. The cause of this last major change in human evolution is obscure, and there is considerable disagreement about its tempo and mode, when and where it happened and what micro evolutionary pattern was followed. With these modern humans we get the first evidence for recognizably modern behavior patterns of many kinds; from cave painting and sculpture, to sophisticated tools and elaborately planned co-operative hunting, to increased population size and density and broadened geographical and ecological range. It is here, finally, that all the truly 'human' characters come together for the first time.

The Neanderthals are recognized as a subspecies of *Homo sapiens* who first appeared around one hundred thousand or less years ago. There seems little doubt that Cro-Magnon man (or what are termed early modern humans) and Neanderthals coexisted on the planet. According to Sarah Bunney (*New Scientist* January 20, 1990), "Skeletal remains provide good evidence of early modern humans in Israel (Qatzeh and Skhul Caves) between 100,000 and 90,000 years ago. By at least 43,000 years ago, populations of modern people had reached southern and central Europe." *New Scientist* issues of November 26, 1987, and February 25, 1988, held that in the Middle East, the two forms of Neanderthal and Cro-Magnon may have coexisted for tens of thousands of years.

According to James Bischoff, from the use of new dating techniques, it appears that early modern people reached the Spanish L'Arbreda Cave in Catalonia and El Castillo Cave in Cantabria around forty thousand years ago (*Journal of Archeological Science* 16, pp. 563, 577). Both caves had previously been used by Neanderthals. Neanderthal tool technology is known as Mousterian, and Cro-Magnon tool technology is known as Aurignacian. Where skeletal evidence is absent, the tool types are taken as evidence of type. Regardless of the academic debate and the absolute accuracy of the dating, this suggests that multiple humanoid species may have existed on the planet prior to the genesis narrative of creation. According to Bunney, "In southwestern France and along the northernmost coast of Spain, as in the Middle East, there are signs that modern people and Neanderthals lived side by side for a time."

The existence of three separate species, without any evidence of intermediary mutants or changes, does not argue that one evolved from the other—indeed, it argues the reverse. The absence of mutant change in *Homo erectus* from 1.5–0.5 million years ago and the emergence of a distinct species 300,000 years ago (in the case of the Neanderthals) and about 35,000–40,000 years ago (in the case of Cro-Magnon), with no mutants either differing or deleterious being found, indicates a stable genetic structure with a secondary structure imposed on the first. However, anthropologists do not accept that Cro-Magnon was a separate species to modern humans but was rather another subcategory.

It would appear that the major factor showing the marked divergence of hominid development was between that of *Homo erectus* and *Homo neanderthalis*. Erectus had a narrow pelvis and reduced braincase. The reduced pelvis indicates a reduced gestation period and hence a smaller, less developed infant and brain capacity. Despite the anthropological argument, there does not appear to be any reason why *Homo erectus* should be viewed as human in any sense other than using sticks and stones for tools, a characteristic also common to apes and sea otters.

The change from *Homo erectus* to *Homo neanderthalis* is an enormous genetic change with the restructuring of gestation, sexual behavior, and familial groupings. Neanderthals had brain capacity equal to modern humans and a better muscled, stronger body. Because of a rounder face for colder climates according to some anthropologists, a case has been developed for reduced frontal lobes. Earlier arguments were for a greater brain size than us, but these may be isolated to some specimens. Arguments concerning restricted vocal cords are reconstructions, and Pilbeam's assertions regarding reduced speech and communication are disputed by anthropologists in the Australian National University.

Neanderthals were late-appearing humanoids separate to and unlike anything preceding them, and contrary to popular evolutionary theory, there is no evidence and no logical reason for accepting that they were the antecedents of the modern humans. Modern DNA testing has revealed that they are utterly unrelated to humans having a twenty-seven-strand DNA system giving us more in common with chimpanzees than them. There is no logically compelling reason why one should accept that *Homo sapiens* or *Homo neanderthalis* evolved from any other species, and neither is there any evidence that indicated that such a construction occurred.

It seems that Dr. Hawkins is correct: the existence of life generates more questions than concrete answers! Life seems to be mysteriously silent about its origin. We would all agree that it is of similar design and that a good case can be made for microevolutionary development. Yet there seems to be an enormous body of evidence to support the claims of objectivists, compatibilists, theists, and literalists—and just as many contradictions. Although this may seem like strike three, in our journey for truth we are not out! Let's continue to seek answers from the experts. If anyone knows, they should.

CHAPTER 4

SEEKING ANSWERS FROM THE EXPERTS

It was no trouble finding expert opinions from every camp. They write books, make videos, record documentaries, and sell products. The creation-evolution controversy fuels them and gives them a platform from which to rail, and rail they do!

The designated spokesman and entrepreneurial genius for the literal, twenty-four-hour day creationist movement is Ken Ham. He is the founder and president of Answers in Genesis, a think tank of scientists and theologians who aggressively defend the six-day position. The organization includes the Creation Museum and a Noah's Ark Adventure near Cincinnati, Ohio, which attract hundreds of thousands of visitors annually; a book-publishing operation; and a series of regional conferences, usually held in large churches. Ham also hosts *Answers … with Ken Ham*, a sixty-second program broadcast daily on radio stations and the internet featuring Ham's commentary on the issues.

Having attended Ham's conferences, I can tell you that his arguments are well received by the conservative Christian community. He is a hero to those who need to be literalists because they believe that any compromise related to the literal interpretation of Genesis is a threat to the literal interpretation of the rest of the Bible. Ham's staff and associates are convincing authors and speakers with eminent qualifications and experience. Much of Ham's work is built on the foundation of renown theologian Henry M. Morris, author of *The Genesis Record*.

Ham believes that the universe was created about six thousand years ago, and that Noah's flood occurred in the year 2348 BC. Much of the dating on which Ham depends is taken from the work of Bishop James Usher (1581–1656), primate of Ireland. Usher developed a complete chronology of human history based on genealogical and historical records from the Bible. Ussher's chronology represented a considerable feat of scholarship: it demanded great depth of learning in what was then known of ancient history, including the rise of the Persians, Greeks, and Romans, as well as expertise in the Bible, biblical languages, astronomy, ancient calendars, and chronology. Ussher's account of historical events for which he had multiple sources other than the Bible is usually in close agreement with other, more modern accounts. For example, he placed the death of Alexander in 323 BC and that of Julius Caesar in 44 BC.

Faced with inconsistent texts of the Torah, each with a different number of years between Flood and Creation, Ussher chose the Masoretic text rather than the Septuagint, which claims an unbroken history of careful transcription stretching back centuries. His choice was confirmed in his own mind because it placed Creation exactly four thousand years before 4 BC, the generally accepted date for the birth of Jesus of the Christian faith. Moreover, he calculated Solomon's temple was completed in the year 3000 BC, so that there were exactly one thousand years from the temple to Jesus, who was supposedly the human fulfillment of the material temple.

Ham believes that the animals carried on Noah's ark produced the biological diversity observed on Earth today. Ham also believes that dinosaurs coexisted with modern humans in the pre-Flood world. He supports his view with scripture and credible evidence on display at the museum. Ham accepts that natural selection can give rise to a number of variations from an original population by Mendelian recombination of already existing genes, but that new genes cannot arise from mutations because this would require adding genetic information. He claims that only intelligence can cause a beneficial mutation and that mutations and natural selection can only remove preexisting information. All of the species are of the same kind, and no new kind can arise from this process.

Ham questions the reliability of radiometric dating, a technique used to date objects such as moon rocks, fossils, and human artifacts. Since 1989, Ham has frequently asked the question, "Were you there?" regarding the origins of life and evolution, implying that knowledge of unwitnessed events can only be inferential and not observational. He continues,

> Creationists and evolutionists, Christians and non-Christians, all have the same evidence—the same facts. Think about it: we all have the same earth, the same fossil layers, the same animals and plants, the same stars—the facts are all the same. The difference is in the way we all interpret the facts. And why do we interpret facts differently? Because we start with different presuppositions; these are things that are assumed to be true without being able to prove them. These then become the basis for other conclusions. All reasoning is based on presuppositions (also called axioms). This becomes especially relevant when dealing with past events.

Many evolutionists and even some creationists take issue with Ken Ham, yet Ham is convinced that science cannot yet prove that God did not create the universe in six days. His immovable position only reinforces the disagreement that exists between naturalists and creationists and even old Earth and young Earth creationists.

Another defender of the creationist viewpoint is Paul Nelson. After receiving his BA in philosophy with a minor in evolutionary biology from the University of Pittsburgh, Nelson gained a PhD in philosophy of biology and evolutionary theory from the University of Chicago. Nelson has been involved in the intelligent design debate internationally for over two decades. He is currently a fellow of the Discovery Institute and an adjunct professor in the Master of Arts Program in Science and Religion at Biola University.

Nelson was an associate of Professor Phillip Johnson of UC-Berkeley, author of the bestselling critique _Darwin on Trial_. Nelson was an organizer of the Mere Creation Conference (1996), where the modern intelligent design research community first formed. Nelson's research interests include the relationship between developmental biology and our knowledge of the history of life, the theory of intelligent design, and the interaction of science and theology. Nelson's scholarly articles have appeared in journals such as _Biology and Philosophy, Zygon, Rhetoric and Public Affairs_, and _Touchstone_, and he has written book chapters in the anthologies _Mere Creation, Signs of Intelligence, Intelligent Design Creationism and Its Critics_, and _Darwin, Design, and Public Education_.

Nelson is a fellow of the Discovery Institute's Center for Science and Culture and a fellow of the International Society for Complexity, Information, and Design. The Discovery Institute is a US nonprofit public policy think tank based in Seattle, Washington, best known for its advocacy of intelligent design. Its Teach the Controversy campaign aims to teach creationist, anti-evolution beliefs in US high schools, positing a scientific controversy exists over these subjects. Thus far, Nelson and his associates have been prohibited from teaching ID creationism in the public schools by the evolutionary scientific community and the courts. In an interview for _Touchstone Magazine_, Nelson said that the main challenge facing the ID community was to "develop a full-fledged theory of biological design."

In the other corner is the bold and unusually aggressive leader of the modern objectivists, Richard Dawkins. An evolutionary biologist, Dawkins is beyond certain that all living things appeared through evolution around four billion years ago. Dawkins came to prominence with his 1976 book _The Selfish Gene_, which popularized the gene-centered view of evolution. In his 2006 book _The God Delusion_, Dawkins contends that a supernatural creator almost certainly does not exist and that religious faith is a delusion—"a fixed false belief" or consonant dissonance, as previously discussed.

In a set of controversies over the mechanisms and interpretation of evolution, in what has been called the Darwin Wars, one faction is often named after Dawkins, and the other faction is named after the late American paleontologist Stephen Jay Gould, reflecting the preeminence of each as a popularizer of the pertinent ideas. In particular, Dawkins and Gould have been prominent commentators in the controversy over sociobiology and evolutionary psychology.

Dawkins is a prominent critic of creationism. He has described the young Earth creationist view that the Earth is only a few thousand years old as "a preposterous, mind-shrinking falsehood," and his 1986 book *The Blind Watchmaker*, contains a sustained critique of the argument from design, an important creationist argument. In the book, Dawkins argues against the watchmaker analogy made famous by the eighteenth-century English theologian William Paley via his book *Natural Theology*, in which Paley argues that just as a watch is too complicated and too functional to have sprung into existence merely by accident, so too must all living things—with their far greater complexity—be purposefully designed.

Dawkins shares the view generally held by objectivists that natural selection is sufficient to explain the apparent functionality and nonrandom complexity of the biological world. In 1986, Dawkins and biologist John Maynard Smith participated in an Oxford Union debate against A. E. Wilder-Smith, a young Earth creationist, and Edgar Andrews, president of the Biblical Creation Society. In general, however, Dawkins has followed the advice of his late colleague Stephen Jay Gould and refused to participate in formal debates with creationists because "what they seek is the oxygen of respectability," and doing so would "give them this oxygen by the mere act of *engaging* with them at all." He suggests that creationists "don't mind being beaten in an argument. What matters is that we give them recognition by bothering to argue with them in public."

In a December 2004 interview with American journalist Bill Moyers, Dawkins said that "among the things that science does know, evolution is about as certain as anything we know." When Moyers questioned him on the use of the word *theory*, Dawkins stated that "evolution has been observed. It's just that it hasn't been observed while it's happening." He added that "it is rather like a detective coming on a murder after the scene … the detective hasn't actually seen the murder take place, of course. But what you do see is a massive clue … huge quantities of circumstantial evidence." Dawkins has ardently opposed the inclusion of intelligent design in science education, describing it as "not a scientific argument at all, but a religious one." He has been referred to in the media as "Darwin's Rottweiler," a reference to English biologist T. H. Huxley, who was known as "Darwin's bulldog."

Much of Dawkins' specific attacks have focused on Michael Behe, one of the leaders of the IDC movement. An American biochemist and author, he currently serves as professor of biochemistry at Lehigh University in Pennsylvania and as a senior fellow of the Discovery Institute's Center for Science and Culture. Behe is best known for his argument for irreducible complexity, which asserts that some biochemical structures are too complex to be adequately explained by known evolutionary mechanisms and are therefore more probably the result of intelligent design. Dawkins strongly contests this argument even though many consider it a major impediment to in evolutionary theory.

Behe has testified in several court cases related to intelligent design, including the court case *Kitzmiller v. Dover Area School District*, which resulted in a ruling that intelligent design was religious in nature. Behe's claims about the irreducible complexity of essential cellular structures have been rejected by the vast majority of the evolutionist scientific community but remains a major plank in the creationist platform.

Behe says he once fully accepted the scientific theory of evolution, but after reading *Evolution: A Theory In Crisis* by Michael Denton, he came to question evolution. Later, Behe came to believe there was evidence at a biochemical level that there were systems that were "irreducibly complex."

These were systems that he thought could not have evolved by natural selection, even in principle, and thus they must have been created by an intelligent designer, which he believed to be the only possible alternative explanation for such complex structures. The logic is very similar to the watchmaker analogy given by William Paley in 1802 as proof of a divine creator.

The books of lawyer Phillip E. Johnson on theistic realism dealt directly with criticism of evolutionary theory and its purported biased "materialist" science and aimed to legitimize the teaching of creationism in schools. In March 1992, a conference at Southern Methodist University brought Behe together with other leading figures into what Johnson later called the "wedge strategy." Following a summer 1995 conference, The Death of Materialism and the Renewal of Culture, the group obtained funding through the Discovery Institute. In 1996, Behe became a senior fellow of the Discovery Institute's Center for the Renewal of Science and Culture (later renamed the Center for Science and Culture), dedicated to promoting intelligent design.

In 1993, Behe wrote a chapter on blood clotting in *Of Pandas and People*, presenting arguments which he later presented in very similar terms in a chapter in his 1996 book *Darwin's Black Box*. Behe later agreed that they were essentially the same when he defended intelligent design at the Dover Trial. Behe's refusal to identify the nature of any proposed intelligent designer frustrates evolutionists, who would love to move him from his scientific box to a religious one, thus making him an easier target.

Francis Collins, project manager for the Human Genome Project, makes a sincere and passionate attempt to insert God into Dawkins' arguments. An evangelical Christian and theistic evolutionist, Collins maintains that life had to have a designer and controlling hand in order to produce the diversity we observe, especially in the case of humankind. He takes particular exception to the idea that life could arise from nonliving matter. "No current hypothesis comes close to explaining how that in the space of a mere 150 million years, the prebiotic environment that existed on planet

Earth gave rise to life." In honesty and humility, Collins refuses to admit what he knows is not true.

To be fair to Collins, it is notable that in the Ben Stein film *Expelled,* under relentless questioning, Richard Dawkins also seemed lost for an explanation for the emergence of life from inanimate matter when he all but admitted that life might have been sown on earth by extraterrestrials. Both Collins and Dawkins, two of the greatest intellects of our time, are honest enough to admit that science's inability to create life in the laboratory creates an intellectual void for evolutionary theory and presents a death blow to proving that life was a cosmic accident from a primordial pool of ooze.

Yet Collins is also a strong critic of creationism. In his book *The Language of God,* he presents a scathing attack upon creationist and IDC leaders by accusing them of manipulating science to fit their preconceived model. He also challenges creationist ideas by saying, "One would expect that that the rank and file of working biologists would also show interest in pursuing these (IDC) issues, especially since a significant number of biologists are believers." This might be true in a neutral environment, but Collins' assertion was put to rest in the movie *Expelled* in which one professional after another was shunned, refused tenure, and was fired for voicing support for exploring creationist scientific ideas. Isn't it amazing that in a society in which 45 percent of people surveyed express belief in a creator, the whole of government institutions, universities, publishers, and the media are bent on discovering, outing, and destroying people who simply want to investigate scientific ideas?

What is amazing is that these "experts" work with the same set of facts but always draw dissimilar scientific conclusions about creation. Behe looks at the irreducible complexity of bacterial flagellum (sort of outboard motors which help bacteria navigate) and discerns that none of the twenty-nine mutations that might have caused the evolution of that appendage were at all beneficial until every one was in place. Dawkins sees a potential benefit in each mutation, regardless of the inability of the flagellum to actually function. Collins sees "fundamentalist Christians" as ignoring and manipulating science to their benefit as he and the rest of the "enlightened evangelicals," who might have been called Gnostics at one time, are the only true guardians of orthodox science. Ham sees the Great Flood as the cause of extensive sedimentary deposits observed in the Earth, whereas Gould sees them gradually forming over millions of years. One side sees the discovery of life on Mars as the ultimate proof of spontaneous generation; the other says that it would prove nothing.

Speaking of Mars, it is the hottest topic in science today. The red planet brings even more disagreement between these experts. Before this book is published, it is anticipated that life may be discovered on Mars. The degree of complexity will be the surprise, but even finding bacteria will be earth-shaking news. Of course, each group will once again find different interpretations in the same facts. Evolutionists are giddy over the possibilities.

If life on Mars is ever proven to exist (or have existed at some point in time), it would mean that the creation of life is not something that happens because of freak chance or divine influence, but is in fact a probable occurrence given the right conditions. Even further, if all that life requires is an aqueous solution like liquid water to grow and thrive (which is the current theory), then the universe is literally teeming with life. The suspected liquid water oceans on some of Jupiter's moons (Europa and Callisto) could be filled with life, and life could still be present underneath the Martian surface, where liquid water and thermal energy are still present. (www.marsnews.com)

The Institute for Creation Research, however, finds the comments of Kenneth Nealson, a NASA microbiologist, comforting and contradictory to the objectivist assumptions.

Furthermore, if scientists were to find clear, unmistakable fossilized bacteria in a meteorite (or soil) from Mars in the future, it doesn't necessarily mean such bacteria evolved from non-life over millions of years. Applying uniformitarian assumptions, secular scientists conclude: "Just as the Mars meteorite ... is thought to have been tainted by Earthly bacteria, samples from Mars, too, may not be what they seem … We think there's about 7 million tons of earth soil sitting on Mars," says Kenneth Nealson, a microbiologist with NASA. "You have to consider the possibility that if we find life on Mars, it could have come from Earth." (*Newsweek*, September 21, 1998, p. 12.

Image Credit: Viking Project, USGS, NASA

NASA

Once again with Mars, there are the same facts but dissimilar conclusions. Missing the step of observation keeps all of our experts in the realm of theory, ever discovering and always interpreting the evidence with a bias toward their own philosophies. There is agreement, but it is divisive agreement, with each camp holding its "own truth" about science and creation. In theology we call this eisegesis, studying the scriptures with a predisposition on what it means. The proper method, exegesis, allows scripture taken in textual, historical, and cultural context to say what it says, free of preconceptions. Make no mistake: scientists, like theologians, are not free from predisposition. Their entire lives form a point of reference from which they cannot escape.

There is no doubt that our so-called experts are experts in defending the particular philosophies that best interpret their views of science. Once again, there are the same facts with multiple conclusions. I don't know about you, but I feel a little frustration kicking in. Might we find the answers in a more thorough study of anthropology?

CHAPTER 5

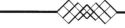

SEEKING ANSWERS FROM ANTHROPOLOGY

For objectivists, orthodox theologians, and evangelical theistic evolutionists, anthropology is the key to determining when modern humankind arrived on the scene. All three viewpoints agree that humans are not dumb animals, operating only on instinct. Humankind is obviously separate and distinct from the rest of animal kingdom. Did we evolve into higher beings, or were we created as higher beings? The answer is a quandary and initiates five other foundational inquiries for the theistic evolutionist.

1. At what point did primates become human beings?
2. How did morality, intellect, judgment, compassion, and spirituality take hold of a particular prehuman species?
3. Was there an actual couple, an Adam and Eve, who made such a transition together simultaneously?
4. How might such human beings have survived among other nonhuman species?

5. Was innocence reality? If so, when and how was it shattered?

These abstract questions do not bother either the objectivist or the creationist. The creationist's version of the direct creation of Adam and Eve by Almighty God avoids such nuance, as does the atheistic or agnostic evolutionist's. It is theistic evolutionists who have a dog in this hunt. It is Rand's curse of eating your cake and having it too. If one believes in a creator and a creation story, one must find a practical way to explain the part of the Genesis narrative that says that the Creator formed man from the dirt with His own hands and pressed His lips to lifeless lips to breathe the breath of life into man.

Do anthropologists and philosophers, most specifically those who support theistic evolution, have answers to these complex questions? You bet they do! Dennis Bonnette, a doctor of philosophy from Notre Dame and an Orthodox Catholic professor at Niagara University, is an avid theistic evolutionist. Following the Church's demand to keep the Gospel message intact, Bonnette has given convincing scholarly answers to these questions in his book *Origin of the Human Species*. It is a laborious and masterfully laid out presentation, which I highly recommend for those who want to delve more deeply into an orthodox alignment and underlying philosophy for theistic evolution. Dr. Francis Collins, in his book, *The Language of God,* likewise makes a similar but less comprehensive philosophical argument.

According to Dr. Bonnette, the evolutionary tree of the twenty-first century is a little less cluttered at the top as mainstream evolutionists and like-minded anthropologists propose that the earliest documented members of the genus *Homo* are *Homo habilis,* the earliest species for which there is positive evidence of use of stone tools, which evolved around 2.3 million years ago. A million or so years later, *Homo erectus*, the first hominid to use controlled fire, arrived on the scene. *Homo sapiens*, anatomically correct modern humans, evolved in Africa possibly from *Homo heidelbergensis* some 50,000 to 100,000 years ago, replacing local populations of *Homo erectus*. People say a picture is worth a thousand words, so here is the creative artwork for what evolutionists propose. The image is a subjective recreation from fragmentary skeletal evidence.

Now, pick the first human. Which one stopped being a brute and became like you and me? Was it the use of tools that pushed humankind beyond the animal barrier? Was it the use of fire? Was it speech? Would the specific moment of intellectual enlightenment have been recognizable and distinguishable? Certainly there were stages of human development that were discernible and measurable: making and using tools, controlling fire, and speech.

Tools are used by many members of the animal kingdom. Otters select rocks from the ocean and river bottoms with which they relentlessly and skillfully crush clams. Chimpanzees use long reeds to extract termites from various environments for food. Tracing the development of tools used by hominoids requires more specific investigation. Evolutionists think that the earliest humanoid tools were made 3.3 million years ago. These prehistoric tools were made from stone, wood, animal bone, antlers, and sinew.

Controlling fire certainly takes more intelligence, more focus, and more skill. All evidence of control of fire during the Lower Paleolithic is uncertain and has at best limited scholarly support. In fact, definitive evidence of controlled use of fire is one of the factors characteristic of the transition from the Lower to the Middle Paleolithic in the period of 400,000 to 200,000 BC.

East African sites, such as Chesowanja near Lake Baringo, Koobi Fora, and Olorgesailie in Kenya, show some possible evidence that fire was utilized by early humans from one million BC. At Chesowanja, archaeologists found red clay sherds. Reheating on these sherds show that the clay must have been heated to 400°C to harden. Koobi Fora sites show evidence of control of fire by *Homo erectus* with the reddening of sediment that can only come from heating at 200–400°C. A "hearth-like depression" exists at a site in Olorgesailie, Kenya. In Gadeb, Ethiopia, fragments of welded tuff that appeared to have been burned were found among created Acheulean artifacts. A site at Bnot Ya'akov Bridge, Israel, has been claimed to show that *Homo erectus* made fires between 790,000 and 690,000 BC.

Other anthropologists state that archeological evidence suggests that cooking fires began in earnest only 250,000 years ago, when ancient hearths, earth ovens, burned animal bones, and flint appear across Europe and the Middle East. Two million years ago, the only sign of fire is burned earth with human remains, which most other anthropologists consider to be mere coincidence rather than evidence of intentional fire. Paleoanthropologist C. Loring Brace agrees with Allman's theory that cooking did not affect development, stating that he has only found evidence of earth-oven cookware from the past 200,000 years, which does not correlate with the earliest known use of fire from 800,000 years ago.

Why is the control of fire important? Wrangham suggested that by cooking meat, it acted as a form of "pre-digestion," allowing less food energy intake to be spent on digesting the tougher proteins such as collagen and complex carbohydrates. Suzana Herculano-Houzel calculated that if we ate only raw, unprocessed food, humans would need to eat for 9.3 hours per day in order to fuel their brains, which use about twice as much resting energy by percentage as other primates. Neurobiologist John Allman states that cooking food contributed to the development of the brain in Neanderthals and early modern humans. Carel van Schaik believes that cooking did contribute to human evolution, and Leslie Aiello suggests that numerous things contributed to the growth of the brain: increased consumption of meat, a smaller digestive system, cooking, and walking upright.

Most anthropologists accept that the first humans made and used tools and controlled fire, but there is one more essential that we must consider: speech. Since the early 1990s, professional linguists, archaeologists, psychologists,anthropologists, and others have attempted to address with new methods what they are beginning to consider "the hardest problem in science." Let's see what they have to say.

Approaches to the origin of language can be divided according to their underlying assumptions. "Continuity theories" are based on the idea that language is so complex that one cannot imagine it simply appearing from nothing in its final form. Thus, it must have evolved from earlier prelinguistic systems among our primate ancestors. Discontinuity theories are based on the opposite idea: that language is a unique trait, so it cannot be compared to anything found among nonhumans and must therefore have appeared fairly suddenly during the course of human evolution. Another contrast is between theories that see language mostly as an innate faculty that is largely genetically encoded, and those that see it as a system that is mainly cultural and learned through social interaction. Evolutionists assume the development of primitive, language-like systems as early as *Homo habilis*, whereas others place the development of symbolic communication only with *Homo erectus* or even *Homo heidelbergensis*, 600,000 years ago, and the development of language proper with *Homo sapiens* less than 200,000 years ago.

Noam Chomsky, a distinguished proponent of discontinuity theory, is a renowned expert concerning the nature of innate universal grammar. He argues that a single chance mutation occurred in a single human about 100,000 years ago, triggering the "instantaneous" emergence of the language in "perfect" or "near-perfect" form, a sort of linguistic punctual equilibrium. Evolutionists assumed that the only change needed was the cognitive ability to construct and process recursive data structures in the mind. From these assertions, should we assume that this instantaneous genetic alteration of the human language faculty is logical? If we do, we can accept Chomsky's theory that language did appear rather suddenly within the course of human evolution.

In 2014, MIT made the following announcement.

> Researchers from MIT and several European universities have shown that the human version of a gene called Foxp2 makes it easier to transform new experiences into routine procedures. When they engineered mice to express humanized Foxp2, the mice learned to run a maze much more quickly than normal mice. The findings suggest that Foxp2 may help humans with a key component of learning language—transforming experiences, such as hearing the word "glass" when we are shown a glass of water, into a nearly automatic association of that word with objects that look and function like glasses, says Ann Graybiel, an MIT Institute Professor, member of MIT's McGovern Institute for Brain Research, and a senior author of the study.

This research seems to confirm Chomsky's assertions, but Professor Suzanne Kemmer proposes that speech may not be as simple as a single mutation. She suggests that speech in humans bridged cognitive, social, and physical dimensions.

Three Dimensions of Development in the History of the Human Species: Neuro-Cognitive, Social, and Physical

From *Inquiry: Critical Thinking across the Disciplines*, Winter, 1996. Vol. XVI, No. 2.

By Linda Elder

Some Preliminary Distinctions

I shall argue that critical thinking cannot successfully direct our beliefs and actions unless it continually assesses not simply our cognitive abilities, but also our feeling or emotion states, as well as our implicit and explicit drives and agendas. I shall argue, in other words, that critical thinking provides the crucial link between intelligence and emotions in the "emotionally intelligent" person. Critical thinking, I believe, is the only plausible vehicle by means of which we could bring intelligence to bear upon our emotional life. It is critical thinking I shall argue, and critical thinking alone, which enables us to take active command of not only our thoughts, but our feelings, emotions, and desires as well. It is critical thinking which provides us with the mental tools needed to explicitly understand how reasoning works, and how those tools can be used to take command of what we think, feel, desire, and do. Through critical thinking, as I understand it, we acquire a means of assessing and upgrading our ability to judge well.

Two Contrary Tendencies of the Human Mind

Every human being enters the world with an initial motivation to have its way and to get what it wants, and thus "naturally" sees the world as designed to cater to its desires. This fact is apparent when we observe the behavior of young children. Their unfailing motto: "It's mine!" As we grow older, we learn methods for getting our way, which are much less blatant and thus less obvious to the untrained eye. Throughout our lives, our own desires and narrow interests are typically in the foreground of our thinking.

As we mature, we learn multiple ways to manipulate others, to influence or control others to get what we want. We even learn how to deceive ourselves as to the egocentrism of our behavior. We have no difficulty coming to conceptualize ourselves as fair-minded, empathetic, kind, generous, thoughtful, and considerate, as concerned, in short, with other persons.

The result is a kind of dualism in us: our selfish, egocentric side, on the one hand, and our capacity to recognize higher values on the other.

Tendencies toward Rationality

Although we often approach the world through irrational, egocentric tendencies, we are also capable, as I have suggested, of developing a "higher" sense of identity. We are capable of becoming non-egocentric people, both intellectually and "morally." Science itself exists only because of the capacity of humans thinking in a non-egocentric fashion--intellectually speaking. Moral concepts, in turn, exist, only because of the human capacity to conceive of responsibilities that by their very nature presuppose a transcendence of a narrow moral egocentrism.

At a minimum, then, I envision the human mind as utilizing its three basic functions (thought, feeling, and desire) as tools of either egocentric or non-egocentric tendencies, both intellectually and morally.

From making fairly sophisticated tools to cooking with fire, to verbal communication, to cognitive understanding, and finally to the big leap to emotional intelligence, we must all agree that the jump from animal to human is huge. Maybe Dr. Bonnette and others were too quick to call

Homo erectus a thoroughly modern human. Maybe it came much later, or in consideration of *Forbidden Archeology*, much sooner. Once again the evidence is conflicting. Yet at some point evolutionists, both theistic and atheistic agree that modern humans had evolved. That begins the real process of understanding from where morality, compassion, spirituality, accountability, and all other uniquely distinctive human characteristics came. Some would call this the soul.

The human soul is a subject deeper than the Laurentian Abyss, so where do we begin? Maybe by looking at humankind's own beliefs about the soul. The soul, in many mythological, religious, philosophical, and psychological traditions, is the incorporeal and (in many conceptions) immortal essence of a person, living thing, or object. According to some religions, including the Abrahamic religions in most of their forms, the immortal human soul is capable of union with the Creator and is unique to humankind. The Catholic theologian Thomas Aquinas attributed soul (*anima*) to all organisms but taught that only human souls are immortal. He further understood the soul to be the first actuality of the living body.

Other religions teach that all biological organisms have souls and that even nonbiological entities such as rivers and mountains possess souls. This latter belief is called animism. The term *soul* can function as a synonym for spirit, mind, psyche, or self. The words *soul* and *psyche* can also be treated synonymously, although *psyche* has more physical connotations, whereas *soul* is connected more closely to spirituality and religion. Although the terms *soul* and *spirit* are sometimes used interchangeably, *soul* may denote a worldlier and less transcendent aspect of a person. According to psychologist James Hillman, *soul* has an affinity for negative thoughts and images, whereas *spirit* seeks to rise above the entanglements of life and death.

Let's examine the soul from the historical perspective.

Plato

Drawing on the words of his teacher Socrates, Plato considered the soul the essence of a person, being that which decides how we behave. He considered this essence to be an incorporeal and eternal. In death, the soul is continually reborn in new bodies. The Platonic soul comprises three parts.

1. the logos, or logistikon (mind, nous, or reason)
2. the thymos, or thumetikon (emotion, or spiritedness, or masculine)
3. the eros, or epithumetikon (appetitive, or desire, or feminine)

Each of these has a function in a balanced, level, and peaceful soul.

Aristotle

Plato's overachieving student and the teacher of Alexander the Great, Aristotle, defined the soul or psyche as the *first actuality* of a naturally organized body. Yet he argued against it having a separate existence from the physical body. In Aristotle's view, the primary activity of a living thing constitutes its soul. When exercised, the various faculties of the soul or psyche, such as nutrition, sensation, movement, and so forth, constitute the "second" actuality, or fulfillment, of the capacity to be alive. A good example is someone who falls asleep, as opposed to someone who falls dead: the former actuality can wake up and go about one's life, whereas the second actuality can no longer do so. Aristotle identified three hierarchical levels of living things: plants, animals, and people. Reason is unique to humans alone.

Eastern Philosophers

Following Aristotle, the eastern philosophers Avicenna Ibn Sina and Ibn al-Nafis further elaborated on the Aristotelian understanding of the soul and developed their own theories on the soul. They both made a distinction between the soul and the spirit, and in particular, the Avicennian doctrine on the nature of the soul was influential among the Scholastics. Some of Avicenna's views on the soul included the idea that the immortality of the soul is a consequence of its nature. In prison, Avicenna wrote his famous floating man thought experiment to demonstrate human self-awareness and the substantiality of the soul. He concluded that the idea of self is not logically dependent on any physical thing and should be considered as a substance. This argument was later refined and simplified by René Descartes when he stated, "I can abstract from the supposition of all external things, but not from the supposition of my own consciousness."

Immanuel Kant

In his discussions of rational psychology, Immanuel Kant identified the soul as the id and stated that the existence of an inner experience can neither be proved nor disproved. "We cannot prove a priori the immateriality of the soul, but rather only so much: that all properties and actions of the soul cannot be cognized from materiality." It is from the id, or soul, that Kant proposes transcendental rationalization, but he cautions that such rationalization can only determine the limits of knowledge if it is to remain practical.

James Hillman

Contemporary psychology is defined as the study of mental processes and behavior. However, the word *psychology* literally means "study of the soul," and psychologist James Hillman, the founder of archetypal psychology, has been credited with "restoring soul to its psychological sense." Although the words *soul* and *spirit* are often viewed as synonyms, Hillman argues that they can refer to antagonistic components of a person. Hillman's psychology is an attempt to give attention to the oft-neglected soul, which Hillman views as the "self-sustaining and imagining substrate" upon which consciousness rests. Hillman described the soul as that "which makes meaning possible, [deepens] events into experiences, is communicated in love, and has a religious concern," as well as "a special relation with death." Hillman takes the Neoplatonic stance that there is a "third, middle position" in which the soul resides.

Contemporary Philosophers

The soul has less explanatory power in a Western worldview. Yet philosophers such as Thomas Nagel and David Chalmers have argued that the correlation between physical brain states and mental states is not strong enough to support identity theory. Nagel argues that no amount of physical data is sufficient to provide the "what it is like" of first-person experience, and Chalmers argues for an "explanatory gap" between functions of the brain and phenomenal experience. On the whole, brain/mind identity theory does poorly in accounting for mental phenomena of qualia and intentionality. Although neuroscience has done much to illuminate the functioning of the brain, much of subjective experience remains mysterious.

Creationists

For the sake of understanding each other's position, let's look at the soul from a creationist perspective, both Christian and Islamic. Most Christians understand the soul as an ontological reality distinct from, yet integrally connected with, the body. Dualists believe that the soul and spirit are one, but there are tripartists who argue that they are distinct and inseparable (Hebrews 4:12). Its characteristics may be described in moral, spiritual, and philosophical terms. According to Orthodox Christian eschatology, when people die, their souls will be judged by God and determined to spend an eternity in heaven or in hell.

Among Christians, there is considerable debate regarding the point at which the fetus acquires a soul and consciousness. This is the reasoning behind the Christian belief that abortion should not be legal. Most Christians regard the soul as the immortal essence of a human, the sea of human will, understanding, and personality; therefore they reason that a fetus is a living human being. Augustine, one of Western Christianity's most influential early Christian thinkers, described the soul as "a special substance, endowed with reason, adapted to rule the body." As mentioned previously, evangelical Christians espouse a trichotomic view of humans (1 Thessalonians 5:23), which characterizes humans as consisting of a body (soma), soul (psyche), and spirit (pneuma).

Philosopher Anthony Quinton states that the soul is a "series of mental states connected by continuity of character and memory, and is the essential constituent of personality. The soul, therefore, is not only logically distinct from any particular human body with which it is associated; it is also what a person is." Souls have sensations and thoughts, desires and beliefs. They perform intentional actions and so are essential. Richard Swinburne, a Christian philosopher of religion at Oxford University, wrote, "It is a frequent criticism of substance dualism that dualists cannot say what souls are."

The origin of the soul has been hotly debated in historical theology. The major theories include soul creationism, traducianism, and preexistence. According to creationism, each individual soul is created directly by God at the moment of conception. Traducianism presupposes that the soul comes from the parents by natural generation, the same as the body. According to the preexistence theory, souls preexist.

The catechism of the Catholic Church defines the soul as "the innermost aspect of humans that which is of greatest value in them, that by which they are most especially in God's image: 'soul' signifies the spiritual principle in man." The Catholic Church teaches that the existence of each individual soul is dependent wholly upon God: "The doctrine of the faith affirms that the spiritual and immortal soul is created immediately by God." Eastern Orthodox and Oriental Orthodox views are somewhat similar in essence to Roman Catholic views, although they are different in specifics.

Protestants generally believe in the soul's existence, but they fall into two major camps about what this means in terms of an afterlife. Some, following John Calvin, believe in the immortality of the soul and conscious existence after death. Others, following Luther, believe in the mortality of the soul and unconscious "sleep" until the resurrection of the dead. Still other Christians reject the idea of the immortality of the soul. They consider the soul to be the life force, which ends in death and will be restored in the resurrection. Theologian Frederick Buechner sums up this position in his 1973 book *Whistling in the Dark*: "We go to our graves as dead as a doornail and are given our lives back again by God, just as we were given them by God in the first place."

Similarly, John Thomas, founder of Biblical Unitarianism, was convinced that we are all created out of the dust of the earth and became living souls once we received the breath of life, based on the Genesis 2 account of humanity's creation. Adam was said to have become a living soul. His body did not contain a soul; rather, his body plus the breath of life were called a soul—in other words, a living being. They believe that we are mortal and that when we die, our breath leaves our bodies, and our bodies return to the soil. They believe that we are mortal until the resurrection from the dead when Christ returns to the earth and grants immortality to the faithful. In the meantime, the dead lay in the earth in the sleep of death until Jesus returns. This belief is sometimes referred to a "soul sleep."

Islam takes a much more mysterious approach to the soul. According to the Quran, Ruh (Spirit) is a command from Allah (God).

> And they ask you, O Muhammad, about the soul (Rûh). Say, "The soul (Rûh) is of the affair of my Lord. And mankind have not been given of knowledge except a little." (Quran 17:85)

Yet like most of Christianity, Islam teaches the soul is immortal and eternal. What a person does is definitely recorded and will be judged at the court of the God.

> It is Allah that takes the souls at death: and those that die not (He takes their souls) during their sleep: those on whom He has passed the Decree of death He keeps back (their souls from returning to their bodies); but the rest He sends (their souls back to their bodies) for a term appointed. Verily in this are Signs for those who contemplate. (Quran 39:42)

It seems that the soul these philosophers (both objectivist and theist) describe is unique to humankind, and therefore it must have an inception. When did the theistic evolutionist's "dumb animal" receive a human soul? According to Dennis Bonnette, the soul arrived in the package that included all the rest of the pieces that made us human rather than animal. This first human, or Adam, is the theistic evolutionist's link to God's plan for the redemption of lost humankind. It is only through the preservation of this vital part of soteriology that Bible-believing Christians can consider theistic evolution.

From this author's historical perspective, anthropology contributes most to the questions of how humankind got here. Yet perplexing issues still remain. Others have faced such challenges, so let's see how these exceptional people made progress in the face of conflicting information.

CHAPTER 6

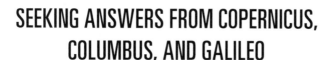

SEEKING ANSWERS FROM COPERNICUS, COLUMBUS, AND GALILEO

During the fifteenth to seventeenth centuries, great scientists, explorers, and thinkers walked where we are walking today, facing the challenges discussed in this book. Without sophisticated equipment or government funding, and facing the Inquisition, they postulated ideas like the orbit of the planets around the sun and the possibility of travel across an undiscovered planet. Even thinking such things was taboo, much less attempting to establish them as truth. They were similar to Michael Cremo and Barry Setterfield, proposing new ideas in a closed-off scientific community that had already made up its mind.

Copernicus

Although the topic was first studied by Hypatia of Alexandria in the fourth century, in 1543 Nicolaus Copernicus, a Polish astronomer and committed Christian, published his treatise *De Revolutionibus Orbium Coelestium* (*On the Revolutions of the Heavenly Spheres*). He dared to present a heliocentric model of the universe. Yet even though his observations and mathematical calculations were correct, it took about two hundred years for the heliocentric model to replace the Ptolemaic model. Thomas Kuhn's comment on the slow response to his obvious genius is relevant to our discussion.

> To describe the innovation initiated by Copernicus as the simple interchange of the position of the earth and sun is to make a molehill out of a promontory in the development of human thought. If Copernicus' proposal had had no consequences outside astronomy, it would have been neither so long delayed nor so strenuously resisted.

As is the case today, Kuhn indicates that the philosophies of that day—Platonism, Roman Catholicism, Judaism, Islam, Deism, and a host of other beliefs—overshadowed the pure scientific discovery, making progress tedious.

Brahe and Kepler

The Danish astronomer Tycho Brahe, who proposed a compromise between the geocentric and the heliocentric theories with his own Tychonic System, contributed to the revolution by showing that the heavenly spheres were at best mathematical devices rather than physical objects. He reasoned that because the great comet of 1577 passed through the spheres of several planets, and the spheres of Mars and the sun passed through each other, their physical nature is subordinate. Brahe and his assistants made numerous and painstaking observations, which allowed Johannes Kepler to derive his laws of planetary motion. Johannes Kepler proposed an alternative model in 1605 (essentially the one finally proven beyond a doubt) in which the planetary orbits were ellipses rather than circles modified by epicycles, as Copernicus used. Kepler's revised heliocentric system gave a far more accurate description of planetary motions than the Ptolemaic one.

Galileo

Starting with his first use of the telescope for astronomical observations in 1610, Galileo Galilee provided support for the Copernican system by observing the phases of Venus and the moons of Jupiter, which showed that the apparently anomalous orbit of the moon in Copernicus's theory was not unique. Galileo also wrote a defense of the heliocentric system, *Dialogue Concerning the Two Chief World Systems*, in 1632, which led to his trial and house arrest by the Inquisition.

Throughout the same period, a number of writers inspired by Copernicus, such as Thomas Digges and Giordano Bruno, argued for an infinite or at least indefinitely extended universe, with other stars as distant suns. Although initially opposed by Copernicus and Kepler, by the middle of the seventeenth century, this became widely accepted partly due to the support of René Descartes.

Newton

The Copernican revolution was completed by Isaac Newton, another outspoken Christian, whose *Philosophiae Naturalis Principia Mathematica* (1687) provided a consistent physical explanation that showed the planets are kept in their orbits by the familiar force of gravity. Newton was able to derive Kepler's laws as good approximations and to get yet more accurate predictions by taking account of the gravitational interaction between the planets. Yet the Roman Catholic Church seemed determined to retain the geocentric model because they saw humankind as the crowning accomplishment of creation and center of the universe.

Columbus

Likewise, Christopher Columbus, known as the boldest navigator of his day, faced a related, equally challenging issue. Was the Earth a flat plane or a sphere? The Roman Catholic Church said it was the former, so could it be successfully navigated? The models of the solar system developed by the Copernicans asserted that the Earth, the sun, and the planets were orbiting spheres. Columbus had observed tall ships sailing to port near his home in Genoa. The tops of the sails always appeared first, followed by the masts and then the deck. Based on this observation, he sailed west, and the rest is history. Unlike the astronomers, he actually risked his life to make a scientific observation.

These brave men and women took risks and faced opposition, as does anyone with a bold new theory. Some were summoned to appear before the Inquisition. It is said that Columbus escaped imprisonment only by summoning an Old Testament scripture to his defense. Why are new ideas suppressed? Because men are stubborn, we cling to old ideas like our comfortable bed shoes. We live in the past and enjoy our comfort zones. But eventually true science triumphs, or at least we hope that it does.

Unfortunately, this is still the case today, and those with their heels dug deep in the sand live on both sides of the evolution/creation issue. Scientists like Dr. Barry Setterfield and Michael Cremo, who have fact-based theories that challenge everything we know about the universe, have not gotten a fair hearing. The conservative men had already begun to formulate their critical responses before the ink dried on their revolutionary theories. Rather than being discussed, their ideas were discarded and hidden. Yet in a hundred years or so, their proposals may prove to be as authentic as those of past heroes.

Let's face it: the biggest obstacle in the cases of Copernicus, Galileo, and Columbus was the scientific ignorance and disinformation of their day. They had to face the archaic belief that the Earth's shape is a plane or disk, and the belief existed long before the Church. Many ancient cultures have had conceptions of a flat Earth, including Greece until the classical period, the Bronze Age and Iron Age civilizations of the Near East until the Hellenistic period, India until the Gupta period, and China until the seventeenth century. It was also typically held in the aboriginal cultures of the Americas, and a flat Earth domed by the firmament in the shape of an inverted bowl is common in prescientific societies.

Jewish acceptance of a flat Earth is found in biblical and postbiblical times. Unfortunately, the Roman Catholic Church bought into the error as well.

Using obscure poetic scriptures not designed to be taken literally, they had made the greatest of hermeneutical errors: taking scripture out of context. Revelation 7:1 is sometimes cited, which speaks of "four angels standing at the four corners of the earth." However, this passage makes reference to the cardinal directions as seen on a compass: north, south, east, and west. Another passage often referred to is Psalm 75:3, "When the earth and all its people quake, it is I who hold its pillars firm."

Today, freshman in a good Bible college understand that there is a difference between biblical poetry and history, parable and narrative. An understanding of context is fundamental, so why did people of much greater learning make such a huge mistake? The answer is obvious to church historians. In fact, this was the principle issue and cause for the Protestant Reformation. The Church was more prone to tradition than scripture.

Thankfully, Columbus, when summoned before the Inquisition, was also armed with scripture. It was another isolated, poetic, anthropomorphic phrase taken from context, but it was scripture. It provided a counterbalance to contemporary thinking. The text is Isaiah 40:22, which states that God "sits above the circle of the Earth." This one thing might have saved Christopher Columbus from being convicted as a heretic.

The fact is there are numerous scriptures that indicate that the Earth is a sphere. These scriptures were penned centuries before Copernicus, some even before Aristarchus, an ancient Greek astronomer and mathematician who presented the first known model that placed the sun at the center of the known universe with the Earth revolving around it. He was influenced by Philolaus of Croton and was able put the other planets in their correct order of distance around the sun. His astronomical ideas were often rejected in favor of the geocentric theories of Aristotle and Ptolemy.

Job 26:7 explains that the earth is suspended in space, the obvious comparison being with the spherical sun and moon. It says, "He stretcheth out the north over the empty place, *and* hangeth the earth upon nothing." Jesus said, "For as Jonah was three days and three nights in the belly of the great fish, so will the Son of Man be three days and three nights in the heart of the earth"—a spherical Earth with a center. Luke 17:34–36 also implies a spherical Earth. Jesus said that at His return, some would be asleep at night while others would be working at daytime activities in the field. This is a clear indication of a spherical, revolving Earth with day and night occurring simultaneously.

Yet from the time of the Romanization of the Catholic or "universal" Church, under Constantine, a new idea began to develop. Whereas the Church had always followed scripture alone (sola scriptura) as a guide for doctrine and practice, now apostolic tradition was taking hold as an alternate guideline. Thus, mainstream Christianity moved away from the perceived simplicity of scripture to something more malleable, changeable, and formational: the habits and beliefs of godly men. For this reason, the leading idea of the Reformation was returning the church to the position of sola scriptura!

The Westminster Confession of Faith, born in Reformation, states,

> VII. All things in Scripture are not alike plain in themselves, nor alike clear unto all; yet those things which are necessary to be known, believed, and observed, for salvation, are so clearly propounded and opened in some place of Scripture or other, that not only the learned, but the unlearned, in a due use of the ordinary means, may attain unto a sufficient understanding of them.

Had this been the case in the fifteenth century, things might have been different. Thus, it was the dogma of Romanism, not the scriptures themselves, that caused the problem science and exploration faced. Nonetheless, the damage was done. Christianity was dealt a body blow from which (in the eyes of some) it can never recover—thus, the derogatory term *flat-earther* for those who believe creationism in any form. The truth is many of the great scientific minds of the sixteenth to nineteenth centuries were devout Christians, including Pascal, Mendel, Newton, Boyle, Pasture, and Kelvin. Each was able to reconcile scripture and science, none accepted a flat Earth position, and their position on the question of Darwinian evolution was skeptical at best.

Looking at how this controversy played out in the sixteenth to seventeenth centuries, is it any wonder that our twenty-first-century clergy struggle with issues that our Christian-scientist heroes of astronomy and exploration from the past faced? This is clearly demonstrated in a recent survey conducted by the Barna Group, which was commissioned by BioLogos and reported on here: https://biologos.org/blogs/archive/a-survey-of-clergy-and-their-views-on-origins (used by permission of BioLogos).

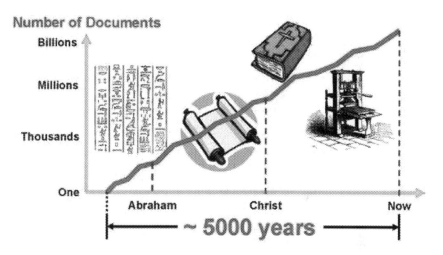

Overall, while a slight majority of the pastors surveyed fall under the label of Young Earth Creationism (54%), sizeable portions of clergy accept Progressive Creation (15%) and Theistic Evolution (18%).

The numbers varied widely based on a number of factors, however. Pastors of mainline churches were most likely to accept Theistic Evolution, while non-Mainline, Charismatic, and Southern Baptist pastors were overwhelmingly Young Earth Creationists. Pastors of larger churches were also more likely to accept Theistic Evolution.

Regionally, the highest percentage of YEC pastors was found in South, while the highest percentage of pastors accepting TE was in the Midwest. Pastors from the western states were the least likely to accept TE.

#2: Most pastors think science and faith questions are important.

Regardless of their views, the majority of pastors surveyed feel that the Church needs to look at how it handles issues of science. 72% of pastors with YEC views and 73% of pastors with TE views agree with the statement that *"the Christian community needs to take a serious look at its understanding of science and human origins in order to maintain its witness in the world."* (The numbers are slightly lower for pastors who hold to Progressive Creation and who are uncertain.)

Similarly, 66% of YEC pastors and 61% of both TE and Progressive Creation pastors agree that *"younger adults today are more concerned than ever about whether faith and science are compatible."*

#3: Clergy think disagreements on science and faith harm our witness (but for different reasons).

Clergy across all three viewpoints feel that disagreements are harming the Church's outreach, but they differ in how they view that harm.

YEC pastors overwhelming agreed (85%) that *"Christian disagreement on matters of creation and evolution is compromising our witness to the world."* However, a majority of TE pastors disagreed with the statement (63%).

Conversely, a majority of TE pastors (63%) agreed that *"The church's posture toward science prevents many non-Christians from accepting Christianity,"* while a majority of YEC and Progressive Creation leaning pastors disagreed (59%).

#4: Pastors aren't avoiding science.

The majority of pastors think that addressing issues of science for their congregations is an important part of their work. Of those surveyed, 72% felt that addressing science issues in the local community was somewhat (51%) or very (21%) urgent. When asked about science on a national and global level, even more pastors felt that addressing science issues is important (43% somewhat and 46% very). Furthermore, 79% of pastors included scientific themes in at least one sermon in the past year, and 40% had included them in at least ten sermons.

The majority of clergy across all four viewpoints also agreed with the statement *"Just as scripture should influence human interpretation of science, science should also inform our understanding of scripture."* The numbers were highest for TE pastors and those who are uncertain (81% and 72%, respectively), though over half of YEC and PC pastors also agreed (52% and 65%, respectively).

Finally, although YEC's are more reluctant than other pastors to say "science should inform understanding of scripture, they strongly agree (84%) that *"The Christian community needs a greater commitment to showing how young earth creationism is consistent with science."*

#5: However, they are concerned about evolution for biblical reasons.

Over half of pastors said they had "major concerns" about the idea that God used evolution. The main reasons for that concern were that the idea "undermines the authority of Scripture" (64%), "views portions of the Bible as non-literal, like Genesis" (62%), "raises doubts about a historical Adam and Eve" (61%), and "raises questions about how and when death and sin entered the world" (59%). However, 26% of pastors saw no concern with the idea that God used evolution.

#6: The majority of clergy accept parts of scripture as symbolic.

60% of the pastors surveyed felt that "some portions of the Bible are symbolic, but all that it teaches is authoritative." Clergy whose views fall under theistic evolution and progressive creation were more likely to accept this statement (79% and 73% respectively), but a sizeable number of YEC pastors (40% among the core followers and 49% among those leaning towards YEC) also agreed with the statement.

#7: Clergy are concerned that changing their views on origins might compromise their ministry.

Over half of pastors (58%) who fell under the YEC category agreed that *"If you publicly admitted your own doubts about human origins, you feel you would have a lot to lose in your ministry."* 41% of pastors in the Progressive Creation group also agreed with the statement. Pastors who were uncertain or who fell under the Theistic Evolution group were less concerned, with only 26% and 17% respectively agreeing with the statement.

"Dazed and confused" best sums up the findings of this research. Many of the clergy are uncertain and even those who believe the Bible won't commit to saying the Genesis narrative is fact, not an allegory. They are closet evolutionists and are careful to avoid the subject of creation in sermons.

The culprit is America's seminaries. As with the rest of academia, the influence in leadership has become largely secularized. Textual criticism is the 21st century toy that mesmerizes our Pastors and religious teachers and keeps them from absolute commitment to the scriptures. The absence of strong support for creationism from the pulpit has energized proponents like Ken Ham to take their message to the church.

This should cause grave concern for creationists and joy for evolutionists. Christian pastors are being negatively influenced and are distancing themselves from a literal understanding the Genesis narrative. Many are not bold enough to declare their newfound faith in naturalism, so they avoid the topic all together. This silence is fueling the ministry and programs of teachers like Ken Ham. His success indicates that Bible-believing Christians are hungry to information on creationism.

Are Setterfield, Behe, Cremo, and Ham quacks, or are they, like Copernicus and Columbus, bold explorers of a mysterious universe filled with unanswered questions?

CHAPTER 7

SEEKING ANSWERS FROM THE SOUL

Which soul? Mine, of course. As a historian, teacher, and author, I must sort this out and inform my readers of the truth. Is the truth found in the previous six chapters? Is it buried beneath experts' bias and manipulated data? I believe that it is. Truth stands up for itself; it is a burning light that shines through the darkest night. Truth shouts from the housetops and mountains; it is not silent and won't be silenced. Before I reveal my findings, let's look at the issues that influenced my reasoning.

1. Evolutionists Have a History of Manipulation

"In the beginning" was the summer of 1926 in the sleepy little town of Dayton, Tennessee. Here, giants on both sides of the creation-evolution issue collided in a cosmic battle for truth. Today, we know the event as the Scopes Monkey Trial. Creationism was represented by three-time democratic, populace presidential candidate William Jennings Bryan. Evolution was represented by the famous agnostic criminal defense attorney

Clarence Darrow. H. L. Menken, the famous American author and cynical reporter for the *Baltimore Sun*, was present to heckle the Christians and cheer on the defense.

The issue might surprise you because in 1926 Tennessee, it was against the law to teach evolution in the public schools. That's right: the Butler Act made it a crime! Now, before we react too harshly, let's remember that it is exactly the reverse today: it is against the law to teach creationism in our public schools. As if by chance, a young substitute teacher, John Scopes, willfully violated Tennessee's Butler Act by teaching evolution. He was charged and held in the town jail, and the issue ignited.

The thing most people do not know is that it was all a setup, organized and financed by the American Civil Liberties Union as a test case in which Scopes had agreed to be tried for violating the hated Butler Act. The teacher of the class, a committed Christian, had refused to cooperate in the scheme. Scopes was charged on May 5, 1925, with teaching evolution from an unauthorized textbook entitled *Civic Biology*, which described the theory of evolution, racial inferiority, and eugenics. Scopes actually encouraged students to testify against him and coached them in their answers.

It was all a cleverly designed plan with an intent and purpose clearly in mind—one that brought about a 180-degree turn in the American public education system. From this point on, our courts were barraged with lawsuits, and that gradually brought us to where we are today. Over the past twenty-five years, IDC proponents and the secularists have attempted to bar each other's information from young minds—minds that should get all of the information available on the issue without prejudice or bias. Why are we afraid to allow students to make a decision on Behe's mousetrap? Is investigating the fact that orbital time and atomic time may vary too much for college-bound students? It seems evolutionists are so fearful some bright mind will find the smoking gun that, like Hitler or Stalin, they must suppress all contrary information.

In the end, Scopes was convicted of a misdemeanor and given a small fine, but the verdict was later overturned on a technicality. The trial gained national attention and produced volumes of literature, including a horribly biased drama, *Inherit the Wind*, considered to be one of the greatest plays (and later screenplays) of the twentieth century. In the play, it seems that the court is overwhelmed by the scientific evidence supporting evolution, but in reality the evidence presented in 1926, mostly based in the theory of embryonic recapitulation, has since been discredited.

The earliest argument proposed by the defense was that there was really no conflict between evolution and the creation account in the Bible as interpreted from the viewpoint of theistic evolution. In support of this claim, Darrow brought in eight experts on evolution, but other than Dr. Maynard Metcalf (a zoologist from Johns Hopkins University), the judge would not allow these experts to testify in person. Instead, they were allowed to submit written statements so that their evidence could be used in the case of an appeal. Dr. Metcalf's testimony is enough for us to make several cogent points regarding these proceedings and the beginning of a conflict that would disenfranchise half of Americans living today who reject evolution. Dr. Metcalf, as would Richard Dawkins or any other evolutionist today, argued from the scientific information that was available to him at the time. Those arguments have been challenged today by both sides and are thought to be inaccurate, racially biased, and unscientific. By the later stages of the trial, Clarence Darrow had largely abandoned the ACLU's original strategy and attacked the literal interpretation of the Genesis narrative and Bryan's limited knowledge of science.

Only when the case went to appeal did the defense return to the original claim that the prosecution was invalid because the law was essentially designed to benefit a particular religious group, which was thought to be unconstitutional. That is where we remain today. The trial revealed a growing chasm in American Christianity and identified two ways of finding truth, one biblical and one scientific. Liberals saw a division between educated, tolerant Christians and narrow-minded, obscurantist Christians. The conventional view was that in the wake of the Scopes trial, a humiliated creationism retreated into the political and cultural background—a viewpoint evidenced in the play and movie *Inherit the Wind* and the majority of contemporary historical accounts. Most objectivists saw the trial as a victory and not a defeat. Yet Bryan's death soon thereafter created a legacy for his defense of scripture and left a leadership void that was eventually filled by other, more qualified Christian apologists. Bryan, unlike the other leaders, brought name recognition, respectability, and the ability to forge a broad-based coalition of fundamentalist and mainline religious groups to argue for the anti-evolutionist position.

Nothing as contrived or blatantly coercive as the Scopes Monkey Trial should stand for truth. Those who support evolution and promote *Inherit the Wind* as great American literature should be ashamed of this fiasco. H. L. Menken, like the media of today, came to Dayton biased against creationism. His articles were scathing, unfounded, and inaccurate. The truth is that this movie has been remade in two parts. The new titles are *Now You See Me* and *Now You Don't*. The Scopes trial itself offers us some insight in our quest for the truth.

2. It Is an Unfair Fight

In today's America, any science supporting creationism is classified as religion, whereas any science supporting evolution is celebrated. There is no fair court of arbitration. This should be true only if evolution is not a religion—which this author believes it is! This was demonstrated in chapter 1: evolution is the religion of the objectivist or secular humanist. I know this is unsettling for those who believe that science is immutable truth and that evolution owns science exclusively, but the fact is evolution is a theory, no matter what Collins, Dawkins, or the Supreme Court says. It should not be protected as exclusive truth, and the science supporting creationism should be allowed in our schools and public institutions. Today's students are bright and well disciplined in their scientific pursuits. Why not let students have access to all the information available and let them decide for themselves? Exposure to outside-the-box radical thinking might give them ideas for radical investigations that could change the world in which we live. Why are evolutionists afraid of a competition among ideas? Any theory that has the exclusive backing of the federal government, media, and public schools and universities is suspect in my book.

3. Evolution Has Been Used to Support Slavery and Racism

Darwin's theory included racial inferiority and unleashed a tidal wave of racial abuse of Africans. Richard Weikart, author of the 2004 book *From Darwin to Hitler: Evolutionary Ethics, Eugenics and Racism in Germany*, states,

> Darwin clearly believed that the struggle for existence among humans would result in racial extermination. In *Descent of Man* he asserted, "At some future period, not very distant as measured by centuries, the civilised races of man will almost certainly exterminate and replace throughout the world the savage (African) races."

Despite the good work that Darwin did in explaining "the survival of the fittest," his thinking on racial inferiority did much harm by reinforcing racial stereotypes. To Darwin, blacks lagged behind whites in evolution. The problem is that this makes perfect sense if one believes in the evolution of modern humans from a variety of independent primate strains. These malevolent statements reveal Darwin's lack of clear understanding of his own theory. Despite efforts by modern objectivists to dispel these inferiority notions, the root of Darwin's tree is despicably corrupt.

4. Evolution Is Constantly Undergoing Change

Evolution has been through hundreds of modifications and revisions since first proposed by Darwin. Even today, many who claim to be objective scientists are embracing radical, complementary theories, such as punctuated equilibrium and alien seeding, to make sense of its flawed science. Neither of these theories is based in true science by any stretch of the imagination. Punctuated equilibrium is the theory that answers creationists' claims that transitional forms are largely absent in the fossil record. Alien seeding deals with the problem of not being able to create life in the laboratory, even with the best equipment and science minds in the world; this failure gives rise to the theory that human beings were placed on earth by aliens. This is science fiction at best. However, when the leading proponent of evolution suggests that either or both may be true, it is time for rational people to question its validity.

Consensus does not establish truth. In the early twentieth century the consensus in America was that the Jim Crow laws were justified. The consensus was admittedly wrong, and thankfully the wrong was eventually set right, but at what cost? The fact that there is a consensus among scientists for evolution means nothing, especially when being a creationist in the academic and scientific community has led to withholding tenure, being censured, and losing one's position. Only scientists who are convinced that they are right would take such risks. When they are silenced by our public universities and institutions, who takes their place? The academic and scientific communities have not genuinely considered contrasting views, especially when it comes to creationism. If you don't believe me, watch Ben Stein's film *Expelled*. Once again, why are evolutionists so fearful of honest debate?

5. Darwin Went Too Far

Darwin's discovery of macroevolution is brilliant and well supported in fact, but extending his theory across species lines without substantive evidence was reckless. That one idea is true does not make the other true as well. For me, even after examining all of the information, the truth is obvious. Using Occam's Razor—"no more assumptions should be made than are necessary"—the idea of a supernatural creator, which requires but one assumption, is more likely than evolution, which requires at least three assumptions.

1. That while using the most sophisticated scientific equipment, laboratories, and supercomputers, scientists cannot produce life in the laboratory, but they assume that it happened by accident in a pool of ooze and in a most inhospitable environment.
2. That in the absence of credible fossil evidence of transitional forms that clearly demonstrate species migration, scientists assume they exist and have even manufactured them or created beautiful

drawings of them, which they sell as truth. In truth, they have no basis in fact. Species do adapt and change demonstratively, but Darwin failed miserable in suggesting that dinosaurs became rodents and monkeys became men.

3. That the mathematical probabilities of events falling into place as evolutionists claim are completely unreasonable. The illustration of a bomb going off in a printing shop and resulting in the production of a dictionary is not as absurd.

My reasoning is based in a systemic failure of the evolutionary community to accept responsibility for the discrepancies in their science and their unmitigated duality in defending those errors with subterfuge and deceit. They have refused to be professional or event honest in their public discourse, choosing to allow the US government and the ACLU to fight their battles. Most have behaved in ways opposite to scientists and scholars.

Understandably, objectivists have no choice but to believe that life began in a pool of ooze, that somewhere transitional forms might exist, and that the absurd mathematical probabilities can be accepted. However, they do not have the right to claim the moral high ground and silence all competing ideas. At the least, they could make the science add up for those of us who still have genuine, reasonable questions and concerns.

Yet this is a battle of public perception, of media and public might over mind. Here is my score on the fight card.

Round 1. Public Perception: Evolution

The media, the US government, and public schools have all imbibed the Kool-Aid. The media has painted creationists as mindless, flat-earth, Bible thumpers. They are despicable deplorables, clinging to their Bibles and religion. Those who dare send their children to government schools allow those students to be shamed and intimidated into compliance. Creationist professors are silenced, shamed, and denied tenure. ID speakers are no longer welcome on college campuses. They are relegated to a dangerous subculture that is bent on taking away the rights of women and the LGBT community.

Even though their scientists have not duplicated life in the laboratory, which would make their theory more palatable, they have, by sheer numbers, created a voluminous amount of narrative and illustration on the subject of evolution. One sees it in every newspaper, magazine, periodical, museum (except Ken Ham's), library, and textbook. It's even on toilet doors. Their so-called evidence is inescapable and overwhelming.

But does volume equate to truth? Did Hitler's "abundant, overwhelming evidence" that the Jews were a "harmful infestation" make it true? Of course not!

Round 2. Science: Evolution

The sheer volume of the information produced by evolutionists and the fact that it is all that one hears or sees in public media is overwhelming. How could so many eminent scientists and our public universities be wrong? Francis Collins, a fellow believer, is convinced of evolution from scientific evidence alone. My visits to the Howard Hughes Institute near my home have brought me into contact with sincere men and women, absent of a spirit of religious bias, who are proponents of evolution based on scientific evidence.

On the other hand, there is something to be said for the simplicity of Behe's mousetrap analogy. The scientific explanations for the Earth not being nearly 13.7 billion years old deserve consideration. I have been to the zoo and have spent lots of time in the woods, and although I find similar design elements humanity shares with the animals, I do not sense that the spirit of humankind could ever have entered a beast through some mystical process that has yet to be unexplained or even made reasonable. The fossil record is also on the side of the creationist, as is the creation of life itself. Creationism lags behind due to limited humanpower and funding, but it seems that it is making up ground.

Round 3. Philosophy: Creationism

Without the benefit of the Bible, Socrates, a nonreligious man, believed in a creator, as did his famous and influential students, Plato and Aristotle.

Even before becoming a Christian, Augustine of Hippo believed in a creator—and it was not that he lacked a reasonable alternative, because the "Christian" Gnostics did not believe that God created the material world. They saw the entire creation as simply a divine hiccup, not an intentional act. Why didn't he join them?

Creationism is good philosophy because it produces beneficial actions. Creationists accept the uniqueness of humankind, the noble expression of divinely initiated human compassion. Therefore creationists build hospitals rather than euthanasia centers, orphanages rather than extermination camps, food banks rather than mortgage banks, rehabilitation centers rather than welfare centers, and pregnancy clinics rather than abortion clinics. Like the former Christian nations of Great Britain, France, Poland, and the United States who opposed atheistic socialism in Germany and Japan, this kinder, gentler philosophy that focuses on helping humanity is appealing. The recent invention of the term *speciesism* (like racism) shows how far the evolutionary community has gone to devalue humanity and human life. Saving human babies seems more philosophically important than saving baby seals.

Round 4. The Experts: Creationism

You can hate Ken Ham if you like, but he is doggedly immovable. On the other hand, Richard Dawkins folds like a cheap kite in a stiff breeze. Over and over again, when pressed hard on the vital question of the origin of life, he has evaded and quibbled. Behe has been challenged and maligned, but he has remained intractable. The creationist belief system is a product of something beyond knowledge, but shouldn't both sides be equally empowered by the truth they embrace? Creationists in China, Africa, and North America are being persecuted for their beliefs, but they remain committed and resolved. Years ago, when the Amish were being imprisoned for not putting their children in government schools, the Supreme Court decided that if individuals were willing to go to prison for what they believe, doing so was not a preference and thus is protected under the First Amendment. When I see that there are academics who are willing to sacrifice tenure and career for endorsing the science of creationism, I am drawn to that level of commitment. If things were reversed, would the evolutionists be as resolute? The evidence indicates that they would not.

Round 5. Awareness: Creationism

One of the proofs for the existence of God is the anthropomorphic argument.

> ".To think at all is to think of God. One cannot think and not think of God, for God is inescapable. The Creator made a creature who is capable of thinking, and when the creature thinks he thinks of the Creator. We cannot do otherwise."

> Written in the universe around man and written on the heart within man is the knowledge of God. This knowledge imparts a sense of accountability, not just a sense of fact. It creates an abiding awareness of personal responsibility to God, not just an intellectual knowledge of the being of God. For man to know of God is for man to know that he is answerable to Him. This is the essence of Paul's argument in Romans, Chapter 1 and it is widely recognized that throughout the world, even in the most remote parts. The creature is desperately seeking the creator. This is the primary reason for atheism; people do not want to be responsible or answerable to a Creator.

In his book *The God Delusion*, Richard Dawkins makes a feeble attempt to what he calls "the God hypothesis," but he devotes only a few sketchy anecdotes to establishing that this God hypothesis is one that has defined religious belief through history or defines it around the world today. He fails to address the vital issues, as is usually the case with arguments he cannot win. I do not know where you are in your journey of exploring the wondrous miracle of creation, but if a still small voice is communicating that you were created with potential, design, and significance, I suggest you listen.

My score card is 3–2 in favor of creationism. What is yours?

The truth is that evolution won by a knockout in round 1 before any other rounds could be contested. So long as evolutionary dogma is favored by the federal government and America's public schools and universities, with the expansive funding attached, creationists have about as much chance of winning the fight as little David with his slingshot against the mighty giant Goliath. And that is the point. With such an advantage, why do half of American's surveyed still favor creationism? They must hear that still small voice that is the anthropological proof that God exists.

CONCLUSION

By now you should have guessed that I favor the traditional creationist viewpoint. Because I am a committed believer and am convinced that the scriptures are inerrant, I have no choice. Yet I am always open and eager to listen to and consider all viewpoints. I enjoy studying alternative theories. Being closed-minded is a waste, and being closed off to new ideas keeps one from progressing toward the truth.

The purpose of this book was not to change minds but to affect hearts. I wanted readers to be exposed to information that is not readily available elsewhere. For me, the jury is still out for the theories of Setterfield and Cremo, but deliberation must continue. These complex topics cannot be reduced to any single argument because one's viewpoint is ultimately determined by one's worldview.

To the objectivist, I wish for spiritual enlightenment, which is the only thing that can change your heart and mind. Might it be possible for you to imagine a beneficent creator? Could you be less condescending to people of faith who do? Like you, they are a product of worldview and are not mindless, evil people. Look for commonality as a bridge to compatibility. You might find that we are approachable and willing to exchange ideas without judgment.

To the young Earth creationists, I implore you to consider that interpreting Yom as an era, rather than a day, does not destroy the veracity of scripture or damage the irrefutable arguments for redemption. If Paul's analogy of cosmic time is applicable to Genesis, the Gospel remains intact. Those who have sound scientific reasons for "millions of years" are not the enemy but your ally. Love them as Christ urged.

To my theistic evolutionary brothers and sisters, I say you can't have your cake and eat it too. The problem with your position has nothing to do with science but everything to do with soteriology. Your theory challenges the vital elements of redemption and leaves humankind hopelessly lost. Dr. Bonnette has made a great start, but more scientific investigation needs to be made if you would reconcile your viewpoint with Jesus's Gospel, as revealed in the book of Romans.

To my traditional old earth creationist friends, I say we must keep open minds, explore all possibilities, and be kind to those with whom we disagree. Young Earth creationists are sincere, and their doctrine is born out of a determination to defend scripture. If Setterfield proves his theories, both sides might be right. Treat objectivists with respect and remember you will never convert them with science but through a chaste life filled with the love of Christ. In 1980, I attended a lecture by a highly placed director at NIH who was a believer. He urged us to n to try to convert agnostics by arguing the virtues of creationism, and he pointed out that once a person becomes a believer, he or she will know the Creator and embrace his omnipotence. Certainly, this was the case with Dr. Collins.

To those who still have an open mind and are not yet convinced of any individual argument regarding creation, I encourage you to keep seeking. You may want to read the books and materials referenced herein. In the end, your decision on this vital issue will affect every aspect of your life. This is not simply a scientific debate but a battle for the soul. The vital cultural and social issues of today are all dependent on the determination of whether man is a cosmic accident and death is the end, or that humankind was intentionally created with intent and purpose and that death is only a door to eternal life.

To all my readers, I wish for us to debate without malice or anger and to seriously consider and respect each other's viewpoints. The great theologian and philosopher Saul of Tarsus wrote, "Let us not look only on our own interests but also on the interest of others" (Philippians 2:4). That's great advice for those of us who want to know the truth about creation.

Printed in the United States
By Bookmasters